UNCHARTED

UNCHARTED

UNCHARTED

HOW SCIENTISTS NAVIGATE
THEIR OWN HEALTH, RESEARCH,
AND EXPERIENCES OF BIAS

Edited by

SKYLAR BAYER

AND

GABI SERRATO MARKS

Columbia University Press   *New York*

Columbia University Press
Publishers Since 1893
New York   Chichester, West Sussex
cup.columbia.edu

Library of Congress Cataloging-in-Publication Data
Names: Bayer, Skylar, author. | Serrato Marks, Gabriela, author.
Title: Uncharted : how scientists navigate their own health, research, and
experiences of bias / edited by Skylar Bayer and Gabi Serrato Marks.
Description: New York : Columbia University Press, [2023] | Includes
bibliographical references and index.
Identifiers: LCCN 2022056979 | ISBN 9780231203623 (hardback) |
ISBN 9780231203630 (trade paperback) | ISBN 9780231555159 (ebook)
Subjects: LCSH: Scientists with disabilities Anecdotes. |
Scientists—Health and hygiene—Anecdotes. |
Science—Vocational guidance.
Classification: LCC Q147 .U563 2023 | DDC 502.3—dc23/eng20230414
LC record available at https://lccn.loc.gov/2022056979

Printed and bound by CPI Group (UK) Ltd, Croydon, CR0 4YY

Cover design: Henry Sene Yee

We dedicate this book to disabled scientists
of the past, present, and future.

CONTENTS

UNCHARTED

UNCHARTED

INTRODUCTION

Charting the Course

SKYLAR BAYER AND GABI SERRATO MARKS

"Charting the course": "Charting" is a nautical term, implying that you are drawing up a route across a nautical chart for your journey. You're creating your own course, not necessarily following someone else's plan. In fact, it may be a unique path forward, especially if you are moving into unfamiliar waters.

No one thinks a PhD will be easy. For many, it involves traveling into unknown metaphorical waters with unexpected professional obstacles. Early in our PhD programs, both of us experienced medical hurdles that we never anticipated. While our medical conditions have challenged us in different ways, we have both noticed that scientific research is rarely designed to accommodate scientists with medical conditions or disabilities, even at the highest levels of success.[1] That could be understandable if we were part of a tiny group, but millions of adults in the United States have a disability—and those figures are from before the COVID-19 pandemic, a mass disabling event.[2]

Disabled researchers remain highly underrepresented in science, technology, engineering, and mathematics (STEM).[3] Scientists with disabilities have so much to offer: we have diverse, creative, and unique ideas that are important for pushing research forward.[4] Nonetheless, we are only able to use our creative skills when we are healthy enough to do research. Both of us have found that we have to be particularly vigilant about making sure that our access to health care and health insurance is consistent and supportive. We have become acutely aware of both the financial and time costs of visiting specialists regularly and the emotional weight of making decisions about our health. We, along with other scientists with disabilities and medical conditions, have succeeded in our careers, but only because we have had access to health care, emotional support, and institutional backing.

Historically, scientists have had to "tough it out" for the sake of research. If we are not tough enough to endure long days in the lab, weeks in the field, or some other metric of strength, we're simply not worthy of being scientists. At least, that's what academic culture tells us. And while those skills and efforts may have been necessary centuries ago, there are plenty of capable scientists who don't fit that mold. Given the prevalence of disabilities and medical conditions among the public—some estimates are as high as one in three adults in U.S. rural areas—it seems strange that our experiences are not shared with the megaphone they deserve.[5] Disabled scientists exist, but our stories haven't become part of academic culture.

The main exception is Stephen Hawking, who is usually the first person to come up if anyone searches for "disabled scientist." But the narrative about his life is almost always about how he "overcame" his "challenges" and was a super-genius. In reality, he succeeded because of the accommodations he received, and nonexceptional scientists should get fair access, too. Moreover, Hawking represents only one kind of disability and one perspective, but there are many other kinds of disability, not all readily visible. If most academics don't know that scientists like us exist, the dominant culture won't change. We want to use our stories to highlight our expertise on this subject and move the conversation forward.

"Uncharted" is a word that came to mind when we were thinking about scientific discovery and journeys of health. A chart is used in navigation, science, and medicine, so to be uncharted is to be off the map, excluded from the data, or not accounted for in health care. Reflecting on our journeys with medical conditions and disabilities in science, we are really navigating our own uncharted journeys in our fields and with our health—creating our own path over a wide expanse of ocean. Often, unbeknownst to us, there are many more of us charting a new path over the same wide expanse. While we all take different paths, we learn similar lessons and tricks to keep our ship afloat and sailing forward. Could we build a stronger community if we came together and shared our journeys with one another?

You'll notice that many of these stories take place out in the field, both on land and at sea, but scientific journeys are not limited to extreme natural environments and outdoor experiences. Researchers don't leave their disability at home when they go to the lab and participate in everyday life, and every type of work can venture into uncharted terrain.

The authorities on life as a scientist with a disability are those who live that reality every day, so we need to be the ones to share our stories and experiences. In this book, we share stories from current *and* former scientists across disciplines, career stages, medical conditions, and disabilities and across other demographics.

While recruiting potential writers, we found very few submissions from people outside the United States or in later career stages. This result reflects a variety of factors: our network, disability legislation in the United States, and changing attitudes around disclosure about health conditions. Haben Girma, the first Deafblind person to graduate from Harvard Law School (and one of Gabi's role models), put it best: "We know that people with disabilities succeed not by magic but from the opportunities afforded by America."[6] We acknowledge our privilege as U.S.-born scientists and hope that this book will open doors for future scientists around the world.

Most of the authors in this anthology are from the United States, but three are from the United Kingdom, one is from South Africa, and one is from New Zealand. Many of our authors are from marginalized communities: they are Black, Native American, Indigenous, Middle Eastern, Asian American, Latinx, multiracial, immigrants, women, first-generation college students, queer, trans, nonbinary, unemployed, veterans, and from many other overlapping groups. After all, disability intersects with all other facets of identity. These additional identities can further complicate the experience of a disabled scientist, as will be very apparent in chapters written by authors with visible marginalized identities. Living with at least one disability or health condition can lead to further marginalization, especially when seeking out diagnoses for separate conditions. We realize that the diversity of our authors is not necessarily apparent throughout our anthology. This was intentional because we did not want authors to feel that they had to share every aspect of their identity, especially if it wasn't relevant to their specific story. Furthermore, there are so few of us that many disabled scientists are easily identifiable by a list of just a few of their identities, even without a name or institution. Several authors also expressed concerns about safety and stigma, including some who chose to remain anonymous. We're extremely proud of the group we've assembled. At the same time, we want to be very clear that this short collection of stories does not represent all disabled scientists. There's a reason that there's no single definition of "disabled"—there's no neat phrase that can encompass all our experiences.

The key questions we ask in this book are: What is it like to be a scientist with a disability or a medical condition? What problems do the writers face every day? Some days? What are the benefits of having a disability? What are the highlights of their life as a scientist, as a human being? How are they supported, and how do they support others? How are their work and their health woven together in their lives? For those authors who have left science, does their story give us clues as to why they left? One of the goals of this book is to show a multitude of human emotions and experiences: sorrow, anger, fear, joy, love, humor, desire, resilience, and curiosity, to name but a few. As a result, many of these stories show the authors as whole human beings and do not focus only on their science or specific research projects. But that doesn't mean they're not scientists—we are all worth more than our work.

In organizing the chapters, one approach would have been to group them by types of medical conditions or disabilities, in which case we would have ended up with sections on mental illness, sensory disabilities, and similar categories. However, when we came up with the concept of *Uncharted*, we wanted to make it clear that our specific conditions or diagnoses are not what define our stories. All of these stories highlight experiences along a journey we were not necessarily expecting, although we all keep moving forward—much as if we were on a challenging nautical voyage. So we divided our anthology into stages of a theoretical nautical journey: "Getting Underway," "Between the Devil and the Deep Blue Sea," "Rallying the Crew," "In the Heart of the Maelstrom," "Reflections in the Water," and "I Am the Captain of My Ship." Each section begins with a description explaining the general themes that tie the stories together.

The diversity within this collection of stories makes it special and accessible to many, including parents, students, teachers, higher education professionals, and anyone with a stake in increasing diversity in science. In short, we hoped to create the type of book that we wish we had, hopefully the first of many more. Most people will need to deal with difficult medical conditions, choices, and stress. Almost everyone is connected to someone with a disability. Therefore, this book is not just for other scientists but for anyone and everyone.

Please enjoy the stories ahead—the authors shared them with generosity, passion, and vulnerability.

NOTES

1. Jessie Shanahan, "Disability Is Not a Disqualification," *Science* 351, no. 6271 (January 2016): 418, doi:10.1126/science.351.6271.418.

2. National Center on Birth Defects and Developmental Disabilities, "Disability Impacts All of Us," Centers for Disease Control and Prevention, September 2020, https://www.cdc.gov/ncbddd/disabilityand health/infographic-disability-impacts-all.html.

3. Colleen Flaherty, "Federal Report Shines Light on Historically Underrepresented Groups in Science," *Inside Higher Ed*, May 2021, https:// www.insidehighered.com/news/2021/05/04/federal-report-shines -light-historically-underrepresented-groups-science.

4. Erica Avery, "Disabled Researchers Are Vital to the Strength of Science," *Scientific American Blog Network*, January 2019, https://blogs .scientificamerican.com/voices/disabled-researchers-are-vital-to-the -strength-of-science1/; Eryn Brown, "Disability Awareness: The Fight for Accessibility," *Nature* 532, no. 7597 (April 2016): 137–39, doi:10.1038 /nj7597-137a.

5. National Center on Birth Defects and Developmental Disabilities, "Prevalence of Disability and Disability Types," Centers for Disease Control and Prevention, October 2021, https://www.cdc.gov/ncbddd /disabilityandhealth/features/disability-prevalence-rural-urban.html.

6. Joseph Shapiro, "She Owes Her Activism to a Brave Mom, the ADA, and Chocolate Cake," NPR, July 2015, https://www.npr.org/sections /goatsandsoda/2015/07/31/428075935/she-owes-her-activism-to-a-brave -mom-the-ada-and-chocolate-cake.

I

GETTING UNDERWAY

*"Getting underway": beginning a trip or journey,
especially in a ship.*

When we are starting out in our careers as scientists, we have transformative experiences that can encourage or discourage us as we consider paths forward. Because we are scientists with disabilities, these formative experiences often intersect with our disabilities. We learn how to move forward with our symptoms or conditions, developing a professional and personal sense of ourselves as we go. Since these experiences are foundational, the first section of this book is dedicated to stories about starting out as a researcher.

Mpho starts this section on his college graduation day, a day that wouldn't be possible without the support of his community. Jenn finds herself in the midst of a medical emergency as a young geology graduation student, midflight to a remote field site. Maureen reflects on how as a blind child she always wanted to be "normal," but she learns in college and graduate school that she can be a scientist with accessible tools, training, and a supportive community. Sami, a neurodivergent scientist, writes a letter of advice and encouragement to her twelve-year-old cousin, Cassy, to persevere in a society that's not made for either of them. And finally, Amanda sets out on her first major oceanographic cruise with big responsibilities, only to discover that her seasickness is just as terrible as it was on small boats—something she didn't anticipate.

These stories are about building skills with a disability or adapting to the new conditions of a research field with a different compass and toolkit than those of nondisabled classmates, teammates, or shipmates. The authors' experiences have helped build their origin stories and propelled them into their future. In these stories, the authors get their metaphorical sea legs as they move forward in their field—or out of it—and in their uncharted journey as disabled scientists.

Mpho Kgoadi

1

ROLLING TO FREEDOM

MPHO KGOADI

Content warning: n/a

Today isn't like any other day. Usually, I wake up with my back tense and aching all over—it must be the way I sleep or rather the uncomfortable single bed from my dormitory. Yes, it must be that. The bed limits my movement during my slumber, and I often wake up numb on my left side. But none of that matters today, which is bright and sunny. I woke up earlier than usual. You see, for the past three years I have studied at this university and endured a lot of hardships with my studies, and today is graduation day—a day to wear my new suit.

This particular suit makes me happy. It is a skinny-fit double-breasted suit in navy blue inspired by one that Cristiano Ronaldo wore at the 2016 FIFA (Fédération Internationale de Football Association) awards. Two weeks ago, I went to see a tailor my friend recommended; he took measurements of my body and told me he would do his "magic." When it arrived, I marveled at his work. Everything was perfect—the fit, size, colour, and stitching; you could tell a lot of attention was given to making this suit. If there were a particular suit I would choose to be my first, this would be it. I received a phone call from my mother telling me that she had arrived at the graduation ceremony.

There was excitement in our voices; after all, I am the first member of my family to be conferred a degree.

My trip to the east campus, which takes about ten minutes, is full of people staring at me. This is my reality. I am used to it, but the staring faces are different today. The faces I usually encounter are ones of wonder? Pity? Amazement? I am not quite sure, but I am sure that these faces are ones of pride as I make my way to the ceremony with my graduation attire. There I see familiar faces that I have seen my whole life—my family. I have never seen this kind of joy beaming from my family; they are so proud of me. They all came to see me: my parents and my three siblings and even my aunt and two cousins are here. This is the happiest I have been in a while. This happiness will not last long, though. As we enter the ceremony, the ushers single me out in a long line of graduates. One of them tells me to follow him backstage because there are stairs leading to the main stage. My wheelchair and I must be isolated from everyone else.

I wasn't always a wheelchair user. This became my reality when I started twelfth grade; however, my condition really started six years before. Growing up, I was very sickly, often skipping school for a week every couple of months after falling ill, usually from fever or nausea. When I wasn't ill, I was very active; you'd never find me indoors because I'd be playing sports. I tried all sorts of sports from karate to cricket to rugby, but I eventually landed on football, mostly because the children around me played football constantly and my older brother forced me to play or watch football with him almost every day. One morning, on our usual walk to school, I found myself lagging behind my friends. I was lagging because I felt a strange tingly sensation in my legs and my feet felt heavy as if someone was holding me down. My friends wondered why I was not keeping up with them, but I just said I was a bit tired, and at that time, I didn't have the words to explain what was happening to my legs. Later that day, I felt weak and woozy, and I was forced to go home early. I knew this was going to be one of those moments when I felt sick and that it would last a week, and I was right.

After spending a week bedridden, I found the strength to get up and join my family for dinner. The moment I stood up, I crumbled to the floor. Was it because I was still weak from being in bed? I didn't think so. Something felt strange around my legs—I could not feel the cold floor on my feet or any pain associated with falling on cold ceramic tiles. I knew something was wrong with me as I shouted for help.

There's a dark cloud at home; the air around the house is sour because another tragedy has struck. I am completely paralyzed from the waist down. The tragedy is not that I have become paralyzed but the fact that I am the second child in the house to be struck by this "curse." My older brother suffered the same fate four years before I did. He collapsed at school while playing football—that happened to be the last time he kicked a ball. Because he was the first to suffer from this mysterious condition, my mother took him to a number of hospitals, doctors, and physiotherapists in search of answers, but nobody knew what had happened or why. At one point the doctors wanted to amputate his legs because they were not sure what the long-term effects of this condition would be, but my mother refused. She hoped that one day her son would walk again. After two years of hospitalization, surgery, and therapy, my brother was discharged. He spent another year at home trying to recover, and he did, partially. He can walk again, using a crutch for assistance.

Nobody could have predicted that a similar fate would befall me, but thankfully, my brother's experience taught us a thing or two. We know what to do, what to eat, which therapist we need to contact, and so forth. My recovery was quite swift but still took over six months. I went to school the following year as a different person, but I felt the same inside. My family and friends and other schoolchildren treated me the same way they did before; I barely noticed that anything was wrong with me— but there was.

As the years passed, my legs failed me to a point where my crutch could not help me balance. I knew there was one inevitable fate if I wanted to maintain my freedom. Before I started twelfth grade, my school donated a wheelchair to me. My family couldn't afford one; they were too expensive. I remember my principal and teacher coming to my house with it; they spoke to my grandmother and she was very grateful, and so was I. You see, this wheelchair would completely change my life. I would never have to worry about which foot to move forward, how to move it, how I should position my crutch to make sure I didn't find myself on the ground, and all the involuntary movements my peers didn't think twice about. This wheelchair meant I could focus on other aspects of my life, such as how to improve my performance at school.

As third-year astronomy majors in university, my peers and I were forced to go to an observatory as part of lab experiments; I was very excited to have a hands-on experience with the telescope. We went to the Hartebeesthoek Radio Astronomy Observatory, in the west of Johannesburg, South Africa. We spent a week learning about radio telescopes and had to perform experiments. On the first day of orientation, we had to climb a ladder atop the twenty-six-metre telescope, and of course ladders and stairs are my nemesis, so I was forced to the sideline as I watched my fellow students and friends ascend to the dish. This was one of many moments when I felt embarrassed, as I was once again reminded of my physical limitations. It would not stop there, as an important aspect of being an astronomer is being at the control room, controlling the telescope and gathering data. I found that the control room was situated on the first floor at the top of a flight of stairs. My friends helped by carrying me and my heavy automatic wheelchair up and down the stairs every day for the whole week. This was embarrassing because I had to watch my peers struggle every day carrying me, but they did not seem to mind, which helped me feel a bit at ease with the whole situation.

There are little constant reminders of my limitations everywhere I go.

The physics building, where I spend most of my days, has only one wheelchair-accessible toilet, which is on the ground floor, and my office is on the second floor. I have to use the back entrance to the building because there are stairs leading to the front door. I once asked for modifications, but I was told that the building cannot be restructured because of "heritage" or something like that. I cannot enter the lecture hall unless both doors are open, and when I do enter, I find myself without a desk, so I often have to write on my lap. I had not realized how bad my eyesight was until I had to attend a lecture and there was no ramp leading to the front. All I could do was listen to the lecturer from the back room and copy my friend's notes as he sat next to me.

So on this special sunny day, I sit behind the curtains on stage eagerly awaiting the dean to say my name. I remember days when my late grandmother took care of me when I was ill, when she told me I was more than capable of achieving my dreams, that everything would eventually work out. I knew that this is what she meant; she was talking about this day.

I hear my name called: "Mpho Kgoadi." As I make my entrance onto the stage to shake hands with the dean and be conferred my degree, I am met with a roar from the audience that sends chills down my spine; this moment feels like all the weight from my shoulders has been lifted. This is a feeling I haven't felt in a very long time. I am not used to the spotlight, so I don't know what to do as I approach the dean. Something tells me to gather the courage and face the cheering crowd, and so I do. I hope that my mother and my family can see the pride in my face as I roll to my freedom.

Jenn Pickering

2

REGAINING CONTROL

JENN PICKERING

Content warning: diet, food

The rows of blue leather seats on either side of the aisle began to fade into the dark splotchy edges of the tunnel that was taking over my field of view. The cabin lights seemed to dim, and my focus shifted toward two flight attendants maneuvering a service cart at the end of the tunnel. I felt a familiar wave of panicked confusion—the initial recognition that I was perspiring, that taking a breath suddenly required effort. It took me by surprise every single time, despite the chronic recurrence of the sensation, leading to a rush of anxiety and a mentally frantic but stumbling slow-motion and physically draining search for something containing a significant amount of sugar.

This time I happened to be buckled into an airplane seat, with my tray table down and bags stowed. I needed carbohydrates to regain control and to stop my body from shutting down. I had stashes of candy and granola bars in my carry-on, but I dreaded the potential embarrassment of fumbling my way out of the seat to retrieve the bag. Even if I could get the tray table up and locked into place despite shaking hands, I didn't think I had the physical strength to safely pull the bag down from the overhead compartment without dropping it on someone sitting nearby.

My mental capacity was fading, and I knew it was a matter of minutes before I would be rendered immobile and an emergency would unfold; I needed to decide how to proceed quickly. I scanned my constricting field of view and my focus landed again on the flight attendants. They were animated in stop motion, but they seemed to be moving toward me; they must have had something on that snack cart that would save me. *Just wait for them to make their way back here, count to ten, focus, breathe in, breathe out.*

As I watched them doling out meal trays, I envisioned what would happen if I stood up to approach them and politely asked for a packet of orange peanut-butter crackers. Would they tell me to take my seat, thinking I was just another rude and impatient passenger? Wait my turn, they'd get to me soon enough, they'd say. This anxiety took over my thoughts, my wrists and neck were already sweating, and my torso was beginning to shake. I crossed my hands over my chest to clutch my upper arms, holding tightly enough that maybe I could keep the shaking from becoming noticeable to my fellow passengers. The service cart was a few rows closer. The flight attendants were chatting breezily with the customers, unaware of my panic.

"Jenn, are you cold? Do you want my blanket?" My advisor, Steve, in the middle seat next to me, held out his thin felt blanket; he'd apparently noticed my shaking. Embarrassed, I said thanks, but I was OK. I was saved from further explanation by the arrival of the food cart, and a flight attendant asked me something about the dinner selection. I didn't have enough concentration to understand her question. "Yes, yes," I said, desperate to get the tray in my hands.

"I'm sorry? Ma'am? Chicken or vegetarian?"

I can't remember how I answered; I can't remember which meal I got. After something like twenty hours of travel, too many time zones, and at least a meal or two skipped or otherwise ill-timed, I was delirious, hypoglycemia (low blood sugar) notwithstanding. I remember fumbling desperately with a brownie wrapped in an impenetrable plastic wrapper, probably cursing at it while somebody or everybody noticed and stared. I vaguely remember Steve or maybe somebody else helping me get to the brownie inside, and I know from experience that I dutifully chewed it like a robot, my mouth dry, no joy in the experience because my taste buds had been cut off by my brain minutes before to preserve more important bodily functions like breathing and circulating blood.

As a type 1 (insulin-dependent, historically referred to as "juvenile") diabetic, I was relatively new to insulin therapy, and this was the first time I'd had a severe hypoglycemic event in transit, on the longest trip I'd ever taken. Although I'd been battling this chronic autoimmune disease for some years, it had remained misdiagnosed until just a few months before. With a new understanding of my disease, I felt that I was finally in control of my body and health. However, in those days my lack of confidence led me to believe that diabetics were a liability to the field of geology, and with aspirations of a future in research, I had decided to keep my diagnosis a secret. I suppose I was concerned that it might affect my chances of getting a job after grad school or, worse, that people might think I had gotten into the program only because I helped the university attain some disability quota. While no one, at the time, actually insinuated that to me, there was an inherent shame that came with being diagnosed with a disability, and that shame hit me hard.

I was a first-year graduate student, embarking on my first field season in Bangladesh to collect samples from the subsurface of the Brahmaputra River. I had big dreams of being a field geologist, studying the rivers of the Himalayas, traipsing across the foreland, and collecting sediment to analyze later in the lab. Knowing that a trek across the globe would be quite a medical burden to manage, I met with my endocrinology team before departing for the field season. Anticipating the conditions in Bangladesh, we decided to change my insulin delivery system from the old-fashioned syringes and glass vials, which I had been carrying around inside a cold lunchbox since my diagnosis, to two types of insulin preloaded into injection "pens," which are a bit more forgiving in terms of fragility and temperature stability.

My medical team also recommended that I keep my target blood sugar a little higher than normal, using the rationale that a month or so of high blood sugar wouldn't immediately kill me, whereas any single severe low-blood-sugar event would likely be debilitating and could potentially be fatal, particularly in the remote landscapes I was headed for. This approach meant that I'd be thirstier than usual and I'd have to deal with some extra exhaustion, maybe some headaches, but it also meant that it would be easier to avoid serious hypoglycemic events. And it would all come down to keeping a strict eating schedule and doing lots of blood-sugar monitoring. The plan was to test my blood sugar with a fingerstick lancet and a glucose monitor about ten or fifteen times each day: upon waking, before and after eating, before and after any physical activity, and before bed. *Sounds great; easy enough; I can control this autoimmune disease.*

Of course, I hadn't understood how difficult it would be in practice to maintain a strict mealtime and insulin regimen while flying to the opposite side of the world with multiple airport layovers over several days of travel. It didn't matter how high my target blood sugar was; the long-acting insulin that I took each day to cover my body's background glucogenesis (the body's natural process of generating sugar) didn't care that I was crossing time zones. If I gave an injection of insulin at seven A.M. in Nashville, it continued lowering my blood glucose for twenty-four hours, whether that twenty-four-hour period ended at four P.M. in the Doha airport or seven P.M. in the Dhaka airport or somewhere in the sky between. All I could do was try my best to stay on top of it, but I had no experience with international travel, and the result was a hypoglycemic event high in the sky, somewhere over the Himalayas, caused by too much insulin and not enough glucose (from the carbohydrates in food).

The brownie was enough to bring me around from my shaking hypoglycemic state—emergency averted (this time), but the recovery is not an instantaneous process. There is usually a fifteen- to thirty-minute window between ingesting carbohydrates and the digestion and movement of that sugar into the bloodstream. For me, this waiting period tends to be calming; the mental anxiety attack begins to fade as my vision widens and the lights begin to brighten and I slowly regain control of my dexterity and coordination.

In contrast, that fifteen-minute waiting period can be nerve-wracking for any bystander watching me go through a hypoglycemic event; my body continues to tremble and my speech continues to sound slurred and labored, my sentences discombobulated. Even if I try to convey that I'm on the path to recovery, it's hard for someone witnessing a minor medical emergency to remain calm, particularly when they are unfamiliar with the experience. Whether the bystander is an arms-length colleague or an intimate friend or relative, their tendency is often to frantically offer me sugary substances. But in my impaired mental state, I can easily end up eating too many carbohydrates to correct my low blood sugar, which results in a new state of high blood sugar, for which I inject a bit more insulin. . . . It's easy to end up riding a blood sugar rollercoaster for an entire day following a hypoglycemic event.

I ended up telling Steve about my diabetes while I regained control of my body; he listened with concern and offered me the rest of his dessert (a true act of compassion!). He wanted to know how to recognize the symptoms of an oncoming hypoglycemic event, particularly because we were embarking on a field campaign. I listed the early signs for him, and we talked about managing our mealtimes and food choices in a way that would ease the burden of controlling my blood sugar. The trip suddenly felt less dangerous to me, and despite my apprehension, it was a great relief to have unloaded some of the mental stress that came with hiding a condition that affects so much of my daily life.

We spent several weeks in the field that season, living largely off a diet of rice and other high-glycemic foods. Steve was helpful in his vigilance, though, frequently asking me how I was doing and paying more attention to the pace of the day. I don't recall exactly how many more hypoglycemic events I had, but I can't imagine that I made it home without another low-blood-sugar episode or two. I'm also certain that my next HbA1c test, which measures an average blood glucose concentration over the preceding couple of months, was higher than is deemed to be healthy. But the friends, cultural connections, and confidence I gained that season doing field research are much more important memories now than the blood glucose readings.

More than a decade after that first trip, I have traveled back and forth to Bangladesh a number of times, including one summer-long visit enrolled in a language immersion program at a local university. Along the way, I've experienced countless blood sugar episodes and related conundrums—how does one respectfully inject insulin into one's abdomen while wearing a long *salwar kameez* when dining with new acquaintances in a Muslim community? Nonetheless, most people are understanding and curious, even helpful, and I've been able to manage the disease adequately enough. Each day is a balancing act, but this lifestyle has been rich with experiences, and every hardship is more than worth it.

Maureen J. Hayden

3

CHANGING TIDES

What Does It Mean to Be Blind?

MAUREEN J. HAYDEN

Content warning: bullying

"**A**ll I want is to be normal." This phrase echoed in my mind as I navigated my primary schooling as a child with a disability.

I definitely felt like I stood out from the crowd. You could usually spot me walking down the school halls or outside corridors: white cane in hand, sunglasses and hat on, and a fanny pack equipped with a handheld magnifier, a monocular (like binoculars but one eyepiece) and other low-vision aids, and finally a backpack that was stuffed to the brim with large-print or braille materials. I was aware of how much space I took up in the classroom, literally. I had a desk at the front with a large closed-circuit television on top—I could put worksheets or books underneath and project them onto the TV screen to be magnified. I also had a second desk equipped with a braille writer and extra storage space for my large-print textbooks. I felt that my peers would judge me for getting special treatment, even though these were the tools that I *needed* to have equal access to the material in school.

Up until the sixth grade I was accompanied in the classroom by a teacher's aide who would assist me with schoolwork. As a result, I found that I related more to adults than to kids my own age. My teachers often reminded me to go play with the other kids at recess rather than talk with my teacher's aides on the playground. I usually had to make the first contact when socializing with other kids. I can't help but wonder if the fact that I was always surrounded by adults created a subconscious barrier that kept my classmates from making that first social contact.

I participated in many "normal" childhood activities, attending pool parties, rollerblading, biking, drawing with chalk on neighborhood sidewalks, and going to the movies. I was good friends with the neighborhood kids, who often asked me to play. I was also invited to join in activities with my peers. Sometimes they asked me; sometimes their parents asked me; or I was invited through group organizations like Girl Scouts.

I got along with many of my peers during these activities, but there was the occasional teasing or bullying. In particular, I remember being referred to as "four-eyes" when I wore glasses. There was also a time when a staff member at an after-school daycare used my white cane in place of a piñata stick without my consent and broke the cane while beating the piñata. At a young age, I was still learning how to process my feelings, but overall, the bullying made me feel sad and lonely.

I was highly aware of my vision in physical education (PE) class. PE was mandatory, but adaptive sports activities and programs were still fairly new. For most activities, I wore sports goggles to protect my eyes and a hat if we were outdoors on the kickball or soccer field. Mostly I was uncomfortable with any activity in PE involving a ball or flying projectile because I don't have depth perception. In middle school and high school, I was allowed to do a substitute activity for PE—I was on the track and cross-country teams and participated in local fitness programs for PE credit.

I had rebellious phases during my primary schooling—I did not want to use the tools and accessible technology that were available to me. This rebellious attitude was my attempt to be more like my peers instead of standing out. I would walk outside without my sunglasses and squint at the bright sunlight. During one phase I wrote all my assignments in tiny handwriting to match the handwriting of other students, only to realize I couldn't go back and read my notes later while studying. I wouldn't use my large-print textbooks or magnification tools and would hold my books up close to read them, often obscuring my face.

All throughout primary school, I was *the blind girl*. But when I got to college, I had the opportunity to redefine how others saw my disability and how I related to my blindness. I was excited for a fresh start. With this change in mindset also came a change in how I approached accessibility. The young girl who used to avoid anything that would differentiate her from her peers started to learn to embrace tools and technologies that would help her achieve her goals.

When I introduced myself to people in college, I realized that they wanted to get to know me for myself—my vision wasn't the first thing they noticed or commented on. Since I was going to a school in the Northeast, it wasn't long before my peers noticed I didn't have an accent, compared to my peers from New Jersey, New York, or Massachusetts, for example.

I began to realize that the methods I used to succeed in the classroom were often beneficial to other students as well, and my accommodations drew questions of curiosity from my peers, not questions of belonging. For example, I prefer high-contrast colors on a whiteboard, and often these colors are beneficial for the rest of the class as well. Black and blue whiteboard markers provide the greatest contrast when presenting written materials to a class. Red and green whiteboard markers might be difficult to see, not only for me but also for any other student in the class who is color-blind. Another example of an accommodation that benefits the class is making presentations using slides with high contrast and large fonts (minimum text size 24) versus slides with photos for backgrounds and excessive text.

In college, I started using my computer more frequently and receiving my notes and presentations from professors in accessible electronic document formats. Accessible documents can be read aloud using technologies like text-to-speech or VoiceOver. In this way, I could view my notes and presentations using the magnifier on my computer and listen to them as well. This improvement may not seem like a big change in today's college climate, but as someone who had always had paper assignments in high school, it took me by surprise. I didn't have to wait for a designated "computer day" in class to use my laptop. Now, instead of having a bulky backpack with eleven-by-seventeen-inch notes, everything I needed was on my computer. I also learned to take advantage of a peer notetaker (provided as part of my accommodations) in my classes to help me make sure I obtained all the information I needed from the lecture.

While some tasks can be assisted by technology, some things have to be learned through practice and time. During my time as a master's student, I took a microbiology lab that was heavy on learning techniques such as streaking plates, performing serial dilutions, performing gram staining, measuring growth curves, performing DNA extractions, and running polymerase chain reaction (PCR) and gel electrophoresis. This class posed a challenge because I had to work with a Bunsen burner and also keep a safe distance from my plate or microbiological culture to keep myself safe and not contaminate the culture. Having no depth perception, I was terrified of using a Bunsen burner to sterilize an inoculating loop. *You want the blind girl to work confidently with an open flame around ethanol?* After orienting myself to my lab bench, receiving feedback and supervision from my peers, and lots and lots of practice, I eventually learned the muscle movements needed to sterilize an inoculating loop. To have learned a typically visual technique through practice and patience made me feel more confident in my ability to accomplish new skills in the lab using nontraditional methods.

One of my favorite accommodations in the lab is using a microscope with a video camera attachment. The video camera can be attached to the microscope through an adapter on the ocular lens or through a third port on the top of the microscope. The camera then connects to a computer monitor with a USB port, and the image under the microscope is projected using a computer program. The projected microscope slide on the computer screen not only allows for a larger, clearer image of the specimen or organism of interest but also allows multiple people to view a microscope slide or specimen at the same time. The lab teaching assistant or professor would often ask to borrow the video microscope to demonstrate key points from the lesson plan or certain specimens to the rest of the class. These video microscope programs have become so sophisticated that you can now take photos, videos, and time-lapse photos and even livestream the session. Video microscopes also prevent eye and neck fatigue and allow for better posture when working at the lab bench.

I learned that even physical activities in underwater environments were possible with accommodations. While pursuing my bachelor's in marine biology at the University of Rhode Island, I had the opportunity to take an open-water scuba certification class. This was my chance to pursue a dream that I'd had since high school and gain valuable experience toward my chosen career path. Before the first day of dive class, I met with the dive safety officer and the dive instructor to discuss how I would scuba-dive. Together we developed several adaptations for diving, such as using a dive computer with a 1-inch number display and backlight and using tactile underwater dive hand signals. I got the idea for tactile communication from reading about Helen Keller in elementary school. Keller would have a person sign into the palm of her hand rather than into the air. The tactile dive hand signals are important for me because I have difficulty seeing objects of similar color, like a black glove against a black wetsuit.

The world looks different under water; colors fade to shades of blue; the turbulence of the ocean determines the clarity of the water and how far you can see in front of you; and everything is slightly magnified so that objects appear larger under water but the underwater world looks just as blurry as the one on land, maybe even more blurry if there is a bad current that leads to poor visibility. My favorite diving method is to go find a spot on a coral reef or rocky intertidal area and perform the sit-and-wait technique. Then instead of spending all of my energy swimming, I sit still and buoyant in the water column and let the marine life come to me. I like night diving best because others have their vision reduced like me. Night divers can only see what their dive light illuminates, so in a way, we are all on the same page under water.

In the last four years, I attended two meetings that changed my relationship to my blindness and the blindness community, and I learned about new technologies through that community. The first was the American Council of the Blind Convention in St. Louis in 2018, and the second was the National Federation of the Blind Convention in Las Vegas in 2019, which I attended as a scholarship recipient. Interacting with my college-age peers and other blind folks, I got to see how they lived full lives and embraced every tool and opportunity given to them. Watching the blind community thrive made me feel supported and reinforced a sense of belonging within a community. The experience made me hopeful for both my professional and personal goals and reminded me that there is always more to learn when it comes to accommodations and accessibility.

At these conventions, I also learned about new technologies. Before 2018, I was not aware of a tool called Braille Screen Input on the iPhone, by which users of VoiceOver can turn the screen into the tactile equivalent of a braille typewriter, using the six braille dots to type instead of a standard keyboard. Now I feel more comfortable using VoiceOver on my phone, and with Braille Screen Input, I have the option to discreetly type messages if needed rather than using dictation. I don't use VoiceOver every day, but it is nice to have the option if the sun is bright outside or if my eyes are tired. Adding VoiceOver and Braille Screen Input to my toolbox has given me more options and flexibility when it comes to how I want to interact with my devices. In the end, finding what accommodation or tool works best in each situation is what is going to make the greatest impact.

I still don't know what "normal" is, but I do know I need to be open to opportunities, open to feedback and constructive criticism, willing to try new things, comfortable with failure, and accepting of change. Our experiences don't have to define us, but they can shape how we approach situations, solve problems, and treat others and ourselves. I don't think there is a "normal." With a little bit of confidence, I don't mind standing out from the crowd.

Sami Chen

4

DEAR CASSY

SAMI CHEN

Content warning: n/a

My name is Sami Chen. I was formally diagnosed as having learning disabilities and dyslexia in my second year of college; the assessment specialist also said that I likely have ADHD. Although several of my family members have dyslexia, ADHD, and/or other labeled learning disabilities, I had a relatively late diagnosis. From an early age, I learned how to overcompensate for things considered "shortcomings" within the standardized U.S. school system.

Despite the social stigma that persists surrounding (dis)abilities, I wouldn't trade being neurodivergent for anything in the world. Being open about my learning disabilities and dyslexia has guided me toward more loving and supportive friendships and deeper familial relationships than I could ever imagine.

Here, I share a letter to my twelve-year-old cousin, Cassandra. In her elementary school days, she opened up to me about her frustrations in English class and with reading assignments, immediately reminding me of when I was her age and tried my best to hide any signs of learning differences. For many years in elementary school, Cassandra spent almost every weekend receiving tutoring, and now she is on the honor roll in middle school. We still talk about our unique and shared experiences navigating the world with dyslexia, and these talks continue to be a special bonding experience. I am incredibly proud of her and wrote this letter for her future guidance. As anyone who has a (dis)ability knows firsthand, there isn't a "magical cure" or sudden "overcoming" that makes our experiences disappear. Nevertheless, we learn to grow into who we are, whether or not the world accepts us.

Dear Cassy,

When you ask us all to wait for a moment, and both our siblings still zoom ahead as soon as they can, I see you and I know what that feels like. Sometimes the world feels like a blur because it's moving like ocean waves but at a faster pace than we can swim, and while we're chasing to catch the last wave, the next one is right behind us. We don't know whether we'll be able to catch the next wave or get swept aside yet again. But in the meantime, we tend to notice the reflection of sunlight on the ocean's surface or find ourselves staring at the most beautiful golden glow of sun shining through the last leaves on the tree each autumn. We can think about each droplet of water being pulled by root hairs and up through a plant's xylem all the way to the furthest leaf that in a few weeks or so you may be helping me collect. This world is so full of wonders that if we were moving at the same wavelength as the rest of the world, we would never get to appreciate these small, beautiful, but significant moments.

School is tough. I wish that our education system had enough funding and support that our teachers weren't overburdened and we didn't have to rely on outdated testing to "prove ourselves" against state standards—standards that seem to forget that even though we may think differently, there's nothing inherently wrong with us. Even if you can't "show" it or won't be rewarded for it in school, the way your brain works is amazing. The nice thing for us is that we started thinking outside the box and had to slowly learn to fit ourselves into the box. This perspective is something we are fortunate to share with our family.

I know that your powerhouse is creativity and innovation. I admire your talented eye for design and your spatial intuition, which shine in how you systematically decorate your room or make very thorough bullet journals. These skills are worth cherishing. As you continue to grow into who you are, you'll begin to appreciate that the way you see the world can actually be a strength. I encourage you to keep pushing and keep advocating for yourself to gather the support you need to follow your dreams.

Whenever you feel that you've hit a roadblock, know that it's OK to ask for help, it's OK to not make everything perfect, and it's OK to not hide when you're struggling. I know that sometimes being honest about our struggles can be challenging. I often struggled to ask questions in classes because I'd equate not understanding a subject with my sense of self-worth and intelligence, which made it very difficult to be open about my academic struggles. Stumbling is a part of the process, and it will help you grow. In a classroom, do not be afraid to ask questions even if you feel uncomfortable.

However, you don't have to be anyone's motivational success story. Although people may approach you hungrily for your story of overcoming hardships and newfound successes, it won't always feel like that. Some days will bleed into weeks or months when you feel you are right back at square one, trying to not get trampled on by expectations, whether they are related to life or work. And when that happens, know that I'll still be right here for you.

While I still struggle with navigating adult life as a grad student and scientist with dyslexia, there have been some perks along the way. I've never felt more joy than from the friendships I've made and maintained as a result of openly sharing my dyslexia and ADHD. Neurodivergence feels like a big, beautiful, rainbow, cotton-candy cloud that we're all wandering around in together. Throughout your lifetime, you will find many "neurodivergent cousins" who just happen to be on the same wavelength as you or who say hello from a parallel wavelength everywhere you go.

I hope that the pure laughter, spontaneity, and creativity that these friendships bring will help you recognize the wonderful nature of being neurodivergent and provide space for you to heal if you need it. I do believe the world is starting to approach learning (dis)abilities with more understanding and compassion. Conversations around being neurodivergent feel more enjoyable now than when I was your age because I feel more aware of the strengths of being dyslexic and having learning (dis)abilities. I now think about friends and family that I love who are also neurodivergent instead of focusing on learning differences between me and my classmates. And like our siblings have had to learn to be patient with us, we too must have patience with our school systems, which have never had a chance to slow down and see the beauty of the water droplets on each blade of grass.

Lastly, now that you've read through all I can think of for now without pulling you along on yet another tangent, here are some words of advice:

On a more practical daily note about reading tips, I'll share some of my favorite fonts and tools. Avenir Next LT Pro and Muli are amazing. They're as legible as Comic Sans without the social stigma. BeelineReader is your friend, and so is anyone who will help you by reading out loud with you. It takes strength to ask for help, and it's important to ask for help often and to identify people who are willing to help you and whom you can also help. Try to take breaks before you feel burnt out, and communicate with your teachers or teammates whenever you feel you're in over your head. As you are weaving together and refining your skills, share them with others. It's important to build reciprocal relationships with people whom you can trust so that you can navigate and thrive in this world together.

Every now and then, take time to pause and reflect on how far you've come. You may be surprised at how many incredible things you've accomplished along the way that leave the rest of us in awe. I am and always have been very proud of you.

With much love and big hugs,
Big Cousin Sami

Amanda Heidt

5

SEA LEGS

Working Around Motion Sickness

Content warning: vomit

I t is August 2016, and I am aboard the *Hi'ialakai*, a former U.S. Navy vessel repurposed to carry out scientific research for the National Oceanic and Atmospheric Administration (NOAA). Two days before, we left Pearl Harbor on an oceanographic research cruise that will keep me and dozens of other crew members at sea off the Hawaiian Islands. For thirty whole days, I will inhabit a shifting world that is never still, never touching my feet to dry land. Even when I am not bolting from task to task, the gentle sway beneath my feet reminds me that we are ever on the move.

This is the first time I have been put in charge of a project with so little supervision, and I have spent months preparing, juggling to-do lists to ready my field gear and myself. Back home, I typed out hundreds of sample labels by hand, the printer spitting out a coil so long it pooled on the floor like arcade tickets. I had to pass a tuberculosis screening, a vision test, and an assessment of my lung capacity. In order to transport my frozen samples with me on the plane, I had to get permission to fly with a dewar, a type of cooler that can hold liquid nitrogen but looks, frankly, like a bomb.

Because of all the effort I've put into planning, and because I've made it to the ship with all my gear, I'm no longer anxious. Rather, I'm socially nervous. The scientists on board come from a dozen different institutions, and most have come in groups. It's just me representing my university, and at first I spend a lot of time alone in the ship's wet lab.

I'm here collecting data for one of my lab's collaborators, taking tissue samples from encrusting invertebrates—sponges, bryozoans, and tunicates—to better understand how they compete with one another for space on bustling coral reefs, where every inch is valuable real estate. It could be boring, I realize. This type of work is somewhat esoteric, and most people, even scientists, don't realize that the colorful slime coating every available surface is alive.

But when you look at these creatures, unassuming though they are, you're witnessing sophisticated warfare on a microscopic level.

When the edges of competing colonies meet, they may begin a battle—releasing digestive chemicals that appear like mustard gas, consuming one another alive. Restricted to no single shape, they may also morph into some new physical form, stretching upward to block out the light and smother their opponents. It's absolute bedlam. The samples I'm collecting will be analyzed for their metabolomes, the compounds and chemicals the animals make as they wage their minuscule battles, to better understand how genetics drive these competitions and shape coral reef diversity.

I hunch over my corner station in the lab, microscalpel poised above the ragged no man's land where a colony of gaudy orange sponge meets another as white as porcelain.

For all my planning, however, there's one hiccup that I should have foreseen, should have planned for but didn't.

From the moment we left land, I've been crushingly, debilitatingly, shockingly seasick. For the first two days, I threw up roughly every half hour, day and night. I worried I might strain a muscle or crack a rib. During the emergency drill, when we had to put on our emergency suits and muster on deck, I crawled up the stairs, and my new shipmates had the good grace to look away as I vomited off the port side. Back in our shared room, the girl in the lower bunk offered me a seasickness patch.

I've started to lose my sight as a rare side effect of that patch, which I hastily slapped onto my neck long after it would have been useful to me, and as a result, I can't see where to cut the sponge colony in front of me. It's not so much a turning out of the lights as a thickening fuzziness, but it is problematic. I make my cuts, but my hands are unsteady, and I'm leaning so near to the table that sometimes my nose dips into the tray of saltwater. Later that evening, entering data in the shared office, I repeatedly zoom in on my laptop until two cells of the spreadsheet take up the whole screen. I quietly return to my room, thankful that ships are required by law to have so many handrails.

When I catch up with my advisor upon my return, he wonders why I agreed to go on the trip in the first place. "That sounds so miserable," he remarks from across his desk, with its comfortable view of the California coast. "Why would you ever put yourself through it?"

It's not an unfair question, and it's also not the first time I've heard it. Motion sickness has been a part of my life since I was a child, and more than a few people have seen it lay me low, though not always in such spectacular fashion. The blindness from the patch is a new side effect.

When he asked me that question, I downplayed the seriousness of what had happened. I didn't want to be left out of future opportunities because of a perceived weakness, real though it is.

I also didn't want to answer truthfully for fear of sounding senti-mental. But I remember thinking that I went on the trip because I am willing to take calculated risks in pursuit of the things that I love.

At each step of my marine biology career, seasickness has forced me to grapple with my passion for fieldwork, a shadow that dogs almost every memory of being on a boat. When I was first being considered for a job as a naturalist with a whale-watching company, for example, I directed the crowd's attention off to one side while covertly losing my lunch off the other. I got the job and held it for a year until I moved to Thailand to teach scuba diving. On that boat—a crisply painted, seventy-foot ship called the *Kiri Marine*—I used to joke to my customers that throwing up was a great way to attract fish. All jokes are rooted in some truth.

I usually feel especially sick on smaller crafts that are subject to the swell, such as the inflatable Zodiacs we used to stage our scuba dives. In theory, the *Hi'ialakai*, a 224-foot vessel, shouldn't have been a problem. It didn't occur to me to take medication before boarding. In fact, I usually avoid antinausea medications because they make me very tired.

But the entire Hawaii trip seemed to contradict my previous experiences with seasickness, making it all the more confound-ing and difficult to manage. For the whole month, I did live off Dramamine and Bonine bartered from my crewmates or pur-chased from the ship's tiny commissary, hoping the side effects would be more manageable (although I avoided the patches—one bout of blindness was enough, thank you). Even though I was tired, I didn't drink coffee for the entire month we were at sea—the most alarming part of this whole story, if you know me personally—because it roiled my stomach. And oddly, I existed in a state of perpetual desperation to get off the ship and onto the smaller dive boats each morning. I never threw up while out on our daily dive trips, and my time under water was often the only time I felt entirely well, with the cool weight of the water settling over me.

While motion sickness isn't a chronic health condition or a disability, it does often incapacitate the afflicted and can occasionally get so bad, as it did in Hawaii, that it presents a real medical risk. Lying in my bunk, I couldn't keep water down, so I sucked on powdered electrolyte mix like it was Pixie Dust. There was talk of sending me back to shore in the days after my poor reaction to the patch, before we got too far from port. I felt like, and probably was, a liability.

But motion sickness is surprisingly common, and it's something many of us have to factor into our work. According to the U.S. National Library of Medicine, as many as one in three people are considered "highly susceptible." It's comforting to know this, much like it's comforting to understand that no problem you experience in life is unique. Navigating through academia over the last decade, I've come to realize just how much scientific progress is carried along on a tide of vomit.

Indeed, it would take several peoples' fingers and toes to count the number of scientists I've witnessed throwing up. The dive safety officer at my master's institution, for example, is perhaps the most seasick person I've ever met, and no underwater research takes place without her permission. And through the door that links our labs, I used to eavesdrop on the girls who go whale spotting in Cessnas, moving in rhythmic, nauseating circles over the sea as they look through peepholes in the floor. I once overheard them comparing notes on which foods tasted best coming back up (it was peanut butter).

Back on the *Hi'ialakai*, after climbing into my bunk on that night when I couldn't see, I told myself that if I woke up the next day still blinded, I would go to the ship's hospital.

Fortunately, taking the patch off cleared my vision overnight, although I continued to be torturously ill for many days. Slowly, however, my equilibrium recalibrated to a point where I could engage more with both people and my work, although I still popped seasickness pills throughout the trip. I kept my food down and even picked up a recipe for fried alligator cutlets from the cook. After a week at sea, I was functioning more or less as I would normally, happy to be there. In an email I wrote to a friend off the island of Moloka'i, I told him that I hadn't felt so energized by my work in a long time.

The *Hi'ialakai* was decommissioned in December 2020 after sixteen years of ferrying NOAA scientists around the world— her third life, actually, following decades in the navy and the coast guard as the *Vindicator*. When I read the news, I was deep in pandemic isolation, feeling far adrift from that time in my life. The previous month, I'd left California to try my hand at desert living, settling into a new version of myself as a science journalist in Moab, Utah. Tucked in boxes in my closet were the field journals from my time in Hawaii, where so many of the details from this story were preserved. I pulled them out, wanting to recapture some of what I'd taken away from that trip five years before.

More than being sick, what I wrote about in those pages were the professional and personal relationships I built and the ownership I learned to take over my work that directly translated into my own thesis research, which would begin the following year (and did not involve boats). I'm still on friendly terms with a few of those scientists who started out as strangers, including one who's getting married this week. I sent her a framed photo of us, wetsuits around our waists and hair stuck to our faces following one of the more memorable dives of the trip.

We'd been given clearance to sample in an area closed to the public because of unexploded military ordnance—we could dive, but we weren't allowed to touch the seafloor. It was an absolutely stunning spot, left alone for many years, and I only saw it because of my work as a scientist. As someone who studies the ocean and grapples with the many issues facing coral reefs worldwide, I felt especially invigorated to see what a reef can look like when humans are removed from the equation.

Each person must decide what level of discomfort they are willing to accept, and I would never urge someone to live in pain for what is, at the end of the day, a job. For me, that choice comes down to a combination of temperament—I love a good suffer-fest—and the goals that I have set for myself while allowing those goals to change. I've since left academia, not because I couldn't stomach another bout of illness or because I was unsuccessful as a scientist, but because I realized how much I enjoy sharing science with others. I hope to one day make it back onto a research cruise as a reporter, there to document important work.

My seasickness, in all its gory glory, is intimately linked to my experiences as a scientist and as a curious human being. I could do without it, but in some ways it has made me a more empathetic person and a more thoughtful researcher. I will never shy away from holding your hair back, and I almost always have a ginger candy in my pocket. Science is neither forgiving of error nor glamorous, and I'd like to think that those who are most successful make a conscious decision to enjoy their work. When I go out to sea, I make sure I see the beauty, mindful that the next moment may not be as pleasant.

II

BETWEEN THE DEVIL AND THE DEEP BLUE SEA

"Between the devil and the deep blue sea": A dilemma or difficult position, stuck between two choices. This phrase may originate from the nautical "devil," which refers to a seam between two planks on a wooden ship.

Although most of this book tries to flip the typical deficit-focused narrative of disability by focusing on the positives and complexities of our lives, our health and wellness can deeply affect our attitude and approach to science. The reality is that some conditions are painful or hard, physically and/or emotionally. Sometimes just getting through the day takes effort. These stories talk about the need to disconnect the body and mind, find coping mechanisms, or just live with uncertainty. *A general warning for this section is that all these stories may be particularly heavy, depressing, or anxiety-inducing for the reader.*

To start this section, Daisy shows us, through her eyes as an autistic person, how she makes an overwhelming commute in London to her safe space in her laboratory. Lauren brings us out to Moab, Utah, where she has traveled hundreds of miles to conduct ecology research, only to find herself inexplicably ill and dangerously alone. An anonymous author shares her struggle to get pregnant and how the associated stress negatively affects her mental health. Skylar is the sole member of her research team who is unable to scuba-dive, so she is left on the surface in an open boat on the coast of Maine, struggling with feelings of inability and uselessness. Finally, we end with Furaha, whose phobias are retriggered and exacerbated after the sudden deaths of a friend and family member, but who continues to seek out hope.

These stories are about the authors' struggles with the uneasiness or unsteadiness of their conditions and about how the ground beneath them never quite settles. It's hard to get used to something that constantly changes—especially when we are not in control. That uncertainty can be particularly challenging when it comes to a diagnosis; as scientists, many of us want to apply the appropriate method (treatment) based on our diagnosis. Without a clear direction, it's easy to get stuck, and sometimes we have long journeys to take before we know where to go. You might notice that many of these authors are not looking for advice, suggestions, or questions like "Have you tried . . . ?" Our lived experiences make us experts on our conditions, even when we don't have all the answers.

Daisy Shearer

6

A SAFE SPACE

DAISY SHEARER

Content warning: n/a

I can do this.

Heading downstairs to the front door, I go through my checklist of things I need and rummage through my rucksack to make sure everything is there. I've done this hundreds of times before, and yet each time I leave the house alone I'm filled with the same anxiety. What if I get overwhelmed and have a meltdown? That is always a possibility, but I know that I have developed techniques to minimise the risk. And at the end of the trip there's a safe space waiting for me. I can't wait to arrive so that I can lose myself in my experimental work.

Pushing my worries aside, I step out onto the pavement and plunge my hands into my pockets. My right hand is greeted with the familiar feeling of my tangle (my favourite fidget toy) and I start to twist it in and out of my fingers, trying to focus on the sensation. The sound of traffic is slightly muffled by my ear defenders, but I'm still acutely aware of everything going on around me. It's a lot to take in. Just focus on the tangle. You can do this.

Approaching the tiny train station in my village, I glance down at my smartwatch to check the time and notice the increase in my heart rate. As expected, my anxiety has increased because I'm taking in more sensory information approaching the bustling station. Checking in with my watch helps me gauge my anxiety level because I have difficulty identifying emotions within myself. The physical confirmation of anxiety alerts me to try to stop the fear from escalating further.

The platform is packed with commuters bound for London. Most of the people waiting for the train are huddled inside the shelter, so I stand further down the platform on a spot where I know I'll probably be right in front of a door when the train pulls into the station. It shouldn't be too long now. I gently rock back and forwards, eyes closed and hand fiddling with my tangle in my pocket and practising deep breathing until I hear the roar of the approaching train.

Just as I thought, the train door stops right in front of me, so I step up onto the train and quickly glance down the carriage. It's packed. I decide to stay standing for my short journey, eyes closed so that I can block out at least one of my senses. This way I reduce the amount of sensory information I'm processing, but I can still feel the anxiety and panic rising as I struggle to regulate my emotions. My heart is pounding and I notice my foot tapping—a self-stimulating behaviour ('stimming') that is quite subtle. I've trained myself to do these kinds of stims in public rather than more obvious stims even though the latter would probably be more effective at helping me regulate. To calm down, I try to focus on the repetitive sound of the train rumbling over the tracks, counting in time with the rhythm of the train.

The train comes to a halt and I'm thrown forwards. Opening my eyes, I can see that most of the passengers have stood up to get off. This is the Reading-bound train, so all the commuters need to change onto the Waterloo train here. I'm jostled out of the train and try desperately to escape the crowd so that I can calm down a bit. No such luck. The train station is incredibly overwhelming at the best of times, let alone in the middle of rush hour. I feel tears pricking in the corner of my eyes and start to hyperventilate. My brain is in overdrive trying to process everything, desperately attempting to predict everyone's movement to make sure I don't bump into anyone and cause an unexpected sensory experience that I know could push me into a meltdown or shutdown. My brain craves certainty and control, so being around so many people can be a challenge unless I'm very focused on my objective.

I've somehow managed to get up the stairs and squeeze myself against the wall to let the crowd pass by. Suddenly, this part of the station is quiet. The crowd has dispersed. But my brain is still in panic mode. From experience, I know that I'm in the "rumbling stage," the first phase of a meltdown or shutdown, and could lose control any moment. I am stimming more intensely and feeling extremely out of control, to the point that I no longer feel present in my body. At this point, I can still avoid disaster if I manage myself and my environment. I need to calm down before I can continue my journey into the lab.

I crouch down, leaning my back against the wall, closing my eyes tight, and pressing my ear defenders against my ears. Construction workers also use ear defenders, but mine are purple and have stickers on them. They block out sounds, and when I press them, they provide a deep pressure that soothes me. After a while, I feel calm again. I cautiously open my eyes and stand back up. Luckily, not many people are around, so there are fewer unpredictable and moving visual stimuli for my brain to process. But I still feel the stares of passersby and of the guard standing by the exit gates. Taking a deep breath, I continue through the ticket barrier and out towards the university campus.

Walking through the car park and to the university campus lake brings me back to myself. The familiar surroundings seem to have rescued me from my spaced-out state. I take exactly the same route to the lab every day. The routine is very soothing, and if I ever have to do it a different way, I can become quite distressed, especially on top of the unavoidable sensory onslaught throughout my journey. I decide to stop by the lake and sit on one of the benches to catch my breath before I enter the busier part of campus. Soon I'll be in my safe space. Not long now. Standing up, I take another deep breath and start walking. Before I know it, I am gently pushing open the glass doors of the Advanced Technology Institute.

I know I can't face any colleagues in this state, so I head immediately to the stairs for the ground floor, turn down the corridor, and walk through the first set of doors with a big sign that says, "Hyper Terahertz Facility." The main optics lab is in front of me at the end of the short corridor, but I'm here for the spintronics lab. Glancing to the right, I look at the wall of laser glasses and notice that there's one pair fewer than when I was last here on Friday. Then I turn to the left and smile as I'm greeted by the familiar signs on the door:

> Warning: Laser hazard
> Designated laser area Class 3b
> Warning: Strong magnetic field

Illuminated in green, the interlock sign reads, "NO HAZARD LASER OFF," so I know it's safe to go into the lab. As I swipe my access card, I can't help but think, "I have such an awesome job." Although I am still in a heightened state of anxiety and sensitivity, my work makes it all worthwhile. The excitement that I might discover something new today is always a thrill and never gets old.

I enter the lab only to be confronted by the glare of too many fluorescent lights. Trying to ignore the burning sensation in my eyes, the flickering, and my static-filled vision, I quickly switch off all but the row furthest away from the door. Much better. As my eyes adjust to the dimmer room, I can already feel the stress start to seep away. The gentle background hum of the equipment sets the scene for the star of the show: the repetitive chugging of the magnet system. Others may hear a cacophony of machinery, but to me the machines seem to be speaking to one another, like an orchestra with an invisible conductor, kept in time by that deep, continuous chuffing of the pump. And when something sounds off-key, I know to check that everything is functioning.

After a quick scan of the room to make sure nothing is particularly out of place, I pull the chair from under the desk nestled in the corner of the lab, plonk myself on the floor under the desk, and raise my knees to my chin. A voice in the back of my head chirps, "You're wasting time. You should be getting on with your experiment." I know this. But I need to recover from the journey first. I've learnt the hard way that trying to work through sensory overload is a recipe for disaster. Looking up to the optical bench, I tune in to the sound of the pump. Gently rocking in time and counting in time in my head is a tried-and-tested stim that helps me self-regulate.

From my spot under the desk, my thoughts turn to the large green cylinder sitting atop the optical bench. I first met Emily (named after a *Thomas the Tank Engine* character by my supervisor because she sounds like a steam engine) properly on the first day of my PhD programme. I'd been introduced in passing by my predecessor when I was applying for the project, but it wasn't until I was faced with actually operating her that I got to know her.

Emily is a split-coil superconducting solenoid with optical access from all four sides—she's basically a huge magnet that you can shoot lasers into. She makes the chuffing noise because the system has to be cryogenically cooled, both to keep the superconducting coils of the magnet below their critical temperature and to allow the sample inside the magnet to be cooled down to temperatures as low as 2 kelvin (around $-271°C$ or $-456°F$). When superconductors are below their critical temperature, they no longer have electrical resistance so a lot of current flows through them. Because electrical current generates a magnetic field, a superconducting magnet can be incredibly strong. Emily can reach magnetic fields up to 7 teslas, which is more than the field strength of most MRI machines. By creating extremely cold environments and applying large magnetic fields and light sources (like lasers), we can investigate interesting quantum mechanical phenomena in various materials.

When I was first faced with operating the system by myself, I was terrified. There were so many knobs and valves and gauges to keep track of. It seemed so complicated that I started catastrophizing about all the ways I could mess things up, convincing myself that I'd somehow wreck everything. But over time I came to love working with Emily. The kinetic aspect of laboratory work with her often became the highlight of my week. Something about it gets me into the flow, and I come up with some of my best ideas while I'm tinkering around in the lab.

By now I realise that my breathing is back to normal and I don't feel overwhelmed anymore. It's time to get on with some experiments. I ease myself onto my feet and head over to the optical bench. Noting down the date in my lab diary, I reach for my sample and a pair of nitrile gloves, preparing to mount the sample into the sample stick and eventually into the centre of the magnet.

Lauren A. White

7

WHEN FIELDWORK
DOESN'T WORK

A Broken Bildungsroman

LAUREN A. WHITE

Content warning: detailed medical descriptions, PTSD, trauma, weight

S taring at the stucco ceiling and listening to the 2014 World Cup blaring from the TV in my dingy motel room in Canyonlands National Park, I tried to pinpoint where exactly in my journey things had gone so horribly wrong. Was it when I had painstakingly packed my car back in Minnesota for this fieldwork? When the eerie green sky threatening tornadoes made sleeping impossible in the Badlands? When I felt dizzy while hiking in Rocky Mountain National Park? Was it when I stayed in the cabin in Moab without air conditioning in temperatures above 100°F? Or was it when I left the hospital in Moab after staying overnight to come here? In retrospect, the exact moment of misfiring did not matter; all this reflection could not change the fact that I was in an isolated location, alone, and very sick.

During the World Cup four years before, I had wandered the empty streets of Buenos Aires during the final match, savoring my independence while living alone in a foreign country. Now, with russet rock formations beckoning through the window, I placidly wondered if I might perhaps die here. Two weeks into my stay, I could still barely eat anything, and standing for any length of time left me dizzy, nauseated, and close to blacking out. I had little to no cell reception, and I was scared of driving myself along the windy roads to the nearest community health center. While I had planned my first field season of graduate school meticulously, it quickly became apparent that I had no backup plan for my health.

When I finished my undergraduate degree in biomedical engineering, I knew that I no longer wanted to spend my time hunched over a computer or a microscope. My childhood dreams of working as a field biologist beckoned me away from this fluorescently illuminated existence. After graduating, I landed a summer field job that I spent driving through prairie grasslands before sunrise, watching pronghorn deer, prairie dogs, elk, and horned lizards; I have never felt so free. I wanted that beautiful, untouchable summer to go on forever, so I went to graduate school.

My goal was to combine ecological theory and fieldwork. The fact that my advisor did not have an established field site was not an insurmountable challenge. I raised grant funding, completed extensive paperwork to be able to handle prairie dogs in the field, purchased supplies, and decided that I would drive by myself to the field site in Utah to meet with other collaborators. My little hatchback was heavy and cumbersome, loaded with beakers, scales, and even a small incubator. During my long drive, my symptoms began innocuously enough: some dizziness, lack of appetite, and headaches. It was easy to ascribe these to stress and altitude.

My symptoms became worse by the time I reached Moab, Utah. Temperatures outside were above 100°F (37°C), and I was rushed to the emergency room in an ambulance with presumed dehydration and heat stress. I was given IV fluids and sent on my way after an overnight stay. Since I had dealt with heat stress before, I planned to spend a few nights in an air-conditioned motel not far from my field site. There was no reason to assume that I would not be fully recovered in a few days. But I didn't recover. I had to rely on complete strangers, like the daughter of the motel owner, to drive me to medical appointments.

With no definitive diagnosis beyond a low-grade fever and an elevated white blood cell count, a nurse practitioner informed me condescendingly during my final appointment that I was clearly having panic attacks, that I should "just relax," and that there was nothing else they could do for me. A seed of doubt entered my consciousness: could all of this be psychosomatic? I agonized over whether to stay and hope for the best, to try to make it back to a major city for additional medical testing, or to attempt to drive back to the East Coast by myself. I played this game of "stay or go" for nearly three weeks while I waited for my symptoms to abate.

In an act of grace that I will never be able to repay, my mother flew out to the nearest airport, hired a taxi to take her nearly four hours to where I was holed up, and helped me drive my overloaded car back to the East Coast. In another act of grace, my taciturn father accompanied me to the doctor's once I arrived home. As a young woman just out of college, I often felt that "there, there, little girl" was the default attitude for my health care interactions. My dad's presence made all the difference in whether I was treated like a hypochondriac. I underwent extensive testing but came back negative for the obvious suspects like hyperthyroidism. Even so, I had persistent arrhythmia, orthostatic intolerance, dizziness, nausea, and not surprisingly, anxiety.

My arrhythmia often felt like a trapped bird fluttering angrily in my chest. Other times, it felt like a manual transmission car switching into gear after several bumpy tries: a revving up, a crunchy pause, and then a resetting. It was always disconcerting but sometimes frightening; anxiety about the arrhythmia could heighten my symptoms, often trapping me in a horrifying feedback loop. Even with medication, my symptoms were severe enough that, if I moved from lying to sitting or sitting to standing too quickly, I risked blacking out. Having to move at all was a monumental effort—like pushing through liquid cornstarch. I lost weight quickly because all I could stomach were saltine crackers. Nearly two months later, I saw an electrophysiologist, who said that he had seen similar patterns of autonomic dysregulation in patients with Lyme disease or in response to certain viruses. It was still an incredibly unsatisfying diagnosis of exclusion. Whatever it was, I could feel my body unraveling. My parents, both veterinarians, later admitted to me that they had been afraid that I might not make it.

As a scientist, I want to reduce things to their component parts. I want to understand the underlying mechanism. So, what am I to do when my body and mind fail to be diagnosed or exist in a predetermined category? How can I make meaning out of an experience that I still do not understand? Beyond losing my field season, I grappled with humbling questions: Who am I beyond my physical body? Who am I if I am no longer a scientist? There was no clear turning point, no clear answers. I stayed on antibiotics and anti-arrhythmia medication for almost six months, and eventually I was able to do a little more each day. My first victories were small: shopping by myself or taking hot yoga classes (which reminded me horribly of Moab). Looking back, I am still unsure whether this progress occurred because my symptoms improved, because I had acclimated to my new reality, or some combination of both. Even so, I struggled with flashbacks, nightmares, night sweats, and hypersensitivity to stimuli. Although I wasn't having mental health issues before, I was certainly displaying several hallmarks of PTSD now. During my recovery, I worked on a review article and then, finally, made the agonizing decision of whether to stay in graduate school. I decided to rework my dissertation. Although I was physically weak, I could lean heavily on my engineering training to answer some similar ecological questions my field work was striving to answer with mathematical modeling instead. Over half a year later, I returned to my graduate program in Minnesota to live alone once more and finish classes.

After turning in my dissertation four years later, I hiked the West Highland Way—a 150-kilometer route between Glasgow and Fort William in the Scottish Highlands. It was my first long-distance hike alone since 2013, when I had hiked right before I started graduate school. During this week of hiking, I felt like I was coming home to my body with a new sense of ownership and confidence in my own resilience.

I wish I could say that the exact nature of my diagnosis did not matter, but I cared immensely about it then, and I still feel ripples of shame about it now. When I first returned to Minnesota, I remember a conversation with the therapist who was trying to help me unpack my own stigma around mental health issues. When she asked me whether it really mattered whether or not it was "all in my head," I burst into tears. It felt horrible to contemplate that my experience hadn't been objectively or measurably "real." It felt even worse to consider that my brain, which was my most treasured part of my identity, had been the architect of my undoing. The worst betrayal of all, of course, was the academic stigma that trivializes mental health and values productivity over personhood. Even as I write this, I am afraid of being judged. Will my colleagues, collaborators, and future employers find me less credible, unreliable, hysterical? Would their regard be different if I had a nicely wrapped, physical diagnosis to share?

Fieldwork stories are a critical currency in ecology circles. But importantly, they are victorious bildungsromans. The protagonist faced with a broken-down field vehicle, dangerous weather, or tropical diseases and secondary infection always comes out on top. Defeat is not an option. You can't share your stories without a resolution—that's not the point. Fieldwork stories highlight that you are the most badass of all: you beat out the elements and you beat out malaria (three times), all the while performing feats of science.

My "victory" has been of the more subtle variety; it certainly doesn't make for a pithy story to be shared at happy hour. But I think I am now more successful as a scientist because of this derailment. My health crisis forced me to establish self-care practices, enforce boundaries, and reevaluate my value system. When I felt that I was dying, I reflected a lot on whether it really mattered more if I had finished publishing a paper or had not seen my family in six months. I promised myself that I would only continue to pursue my PhD if I could stay healthy and at least moderately happy. I petitioned to work remotely from the East Coast to write my dissertation. I worked during set hours. I prioritized time with my husband, family, and friends and pursued activities that brought me joy. I still love science, but I love it from afar: it exists safely apart from the most vulnerable parts of me and will carry on just fine without me, even if I were never to publish another paper or conduct another experiment. These days, I am not afraid of dying; the thought of my molecules dispersing and returning to the Earth, sky, and green, growing things fills me with peace. It is the suffering in between that scares me.

We are afraid to admit that sometimes hard work, persistence, and grit are not enough. And yet, without these trials by fire in the field, how can we truly know who has the mettle to make it in science? And if we fail, how can we have faith in ourselves to survive the rigors of academia? Most importantly, what if academia is not a meritocracy to begin with? (Hint: it isn't.) In life, as in statistics, systemic biases plus random effects can equal an individual's perceived failure. But it isn't a failure of the individual; it is a failure of systems that uphold prejudices, stigmatization, ableism, and a narrative of "pull yourself up by your bootstraps." This narrative punishes us all by excluding the voices we need to hear the most in order to grow, to change, and to do better science.

Pill bottle

8

BIRDS, BEES, AND ANXIETIES

ANONYMOUS 1

Content warning: pregnancy and fertility, medical procedure

"**A**nd what brings you here today?" I did not want to be there, in that room with the uncomfortable chairs and degrees from prestigious universities on the walls. I had started and stopped working with a half-dozen mental health practitioners, aggravated with how long it would take to get my mental illnesses managed, and the $40 copays were adding up. "I'm ready to have a baby, and I'd like to get my disorders under control so I have a healthy and happy pregnancy." Simple, right? We met every two weeks to identify thoughts, emotions, and responses; learn coping mechanisms; and gauge progress. I hid the mental health booklets in my bag, ashamed that someone might glimpse the words "depression," "anxiety," "obsessive-compulsive," or "anorexia" and associate them with me. I closed the door before pulling the workbooks out, finding my favorite purple pen, and scribbling down my responses. I was proud, giving myself an imaginary gold star sticker whenever the therapist proclaimed, "Nice work." But as the visits went by, any advances in my mental health felt over-shadowed by one problem—I wasn't getting pregnant.

I lost track of how many times I squinted at a pregnancy test—*is that a line?*—how many cocktails I refused—*you never know*—how many plans I postponed—*I might be taking care of a two-month-old by then.* When my period did arrive, it felt like a punishment, and I thought back to every sit-up I had done, every time I was hit with car exhaust, and the many other things that likely did not affect my inability to conceive but at the time felt like some sort of explanation.

A year and a half later, my partner and I found ourselves in a small, sterile office, staring at a uterus model and having ovulation explained to us by the reproductive endocrinologist. To help with some abbreviations I will use, here are a couple definitions. IUI, or intrauterine insemination, is a procedure during which a catheter is inserted into the uterus and a semen sample is placed. IVF, short for "in vitro fertilization," is an artificial egg fertilization process that is preceded by follicle stimulation and a brief surgical procedure for egg retrieval. The fertilized egg is then transferred into the carrier. Both IUI and IVF involve numerous monitoring visits to ensure that hormones and follicles are at the appropriate stages.

It felt strange. For two weeks, I had regular visits, transvaginal ultrasounds, bloodwork, and check-ins with the office. Then for two weeks, I waited. *Want to test early? Well, I triggered my ovulation with choriogonadotropin (the "pregnancy hormone"), so I'm likely to have a false positive because it stays in the system for a while. Tested anyway, and it's negative? It may be too early for hormone levels to be picked up on a home pregnancy test.*

My psychiatrist appointments were soon consumed by the baby that did not exist. I can't even remember why I started seeing him, but each visit was a laundry list of fears and things I wanted to work through or prevent. Having struggled with an eating disorder for most of my adult life, I was worried that it would reemerge during a pregnancy. *What if I get postpartum depression or psychosis? What if I die during childbirth? What if the baby has a fatal disorder?* I refused medications and filled out mood diaries and whatever forms were handed to me, thinking I could treat this process like any other work: if I put in some extra nights and weekends, I could get ahead. I could get better faster. My therapist was supportive, but simply catching him up on everything that happened since the previous visit left little time for me.

I spent months of "two-week waits" reading blogs and forum posts about trying to conceive, overthinking everything my body did. *Was that an "implantation bleed"? Are these cramps the "tugging feeling" associated with early pregnancy?* I reached for my statistics books to calculate the odds of not getting pregnant after each cycle, factoring in sperm count and all the other research I could find. With each cycle, the odds got smaller and smaller, and I felt worse and worse. *Either I'm a statistical anomaly, or something is wrong.* I started to feel paranoid, thinking that maybe the clinic did not want me to get pregnant and was sabotaging me. *Is it because I have a mental illness? Am I unfit to breed?* I also received no answers as to what might be causing the infertility. *Was I too stressed? Was I stressing about being stressed? What if I made the sperm fall out that time I sneezed after the insemination?* I was desperate.

Anytime I set a boundary during this process, I ended up pushing it. I saw the pictures of my friends and family members with growing bellies and smiling babies. My own clock ticked harder until it was all I could hear. "I won't take meds" became "I won't go higher than this dose," then "I'll only try IUI up to eight times," to "I won't do IVF" and "I'll only do IVF once." By the time I began the IVF process, I could not even think for myself. *Just tell me what to do, and hopefully I don't mess this up too much.* Day after day, I watched and felt my follicles grow. Night after night, I sat in my semiclean bathroom space, psyching myself up to repeatedly inject myself with various concoctions. I was quickly running out of space on my belly and found myself sobbing the last few nights, unsure if I would ever find a spot to insert the needle. Before bed, I looked down and saw the little red and purple dots, the bruising, and some welts from injection site reactions. I wanted to snap a picture and share it with the world, in hopes that others would see it and know they were not alone. To this day, it feels strange that people can share so many things about themselves with complete strangers, but anything other than fertility and positive pregnancy outcomes is kept quiet.

The egg retrieval itself went well. One by one, the technicians, nurses, surgeon, and anesthesiologist came by to explain what was going to happen. *Little one, do you know how many people it took to bring you into this world?* I looked down at the hospital socks and thought the color looked nice against my tanned skin. I counted backwards from a hundred, though I cannot remember how far I got. They were able to retrieve five mature eggs (apparently not every follicle has an egg), and four fertilized normally.

The recovery was unexpectedly painful, and it took everything in me to not cry as I faced the stairs I would have to climb to get to my room. I was surrounded by many comforting objects: heat pad, electric blanket, multiple pillows of various sizes, and a steady supply of tea. My partner would help me get situated, but I always felt the pangs from deep inside and screamed.

I thought about my fertilized eggs as they developed over the week, growing in their dishes, not unlike the cells I care for as part of my own work. I thought about how the cells I work with also belonged to someone and, by some mutation, were able to keep dividing. *Mutation . . . should I have done the genetic testing? It was so expensive.* It is a difficult decision to make. Should I pay thousands of dollars to test embryos ahead of time or carry a fetus through the first trimester to find out then? I felt that whoever made that decision had their company's profits in mind rather than the turmoil of someone who wants to be a parent.

Four embryos seemed like a lot, and I never considered having that many children. I knew they would not all result in live births, but with my luck (statistical anomaly, right?), who knew what could happen? Days went by. A friend messaged me asking if I had an update. I excitedly texted back that I was not sure, but I decided that the number of embryos that would survive would be the number I had transferred. I made a joke about how the next time I was ready for a baby I would just thaw out one of my leftovers. I thought about the embryo storage fees, the age difference for my future children, and the toll that pregnancy and delivery would take on my body.

An email notification appeared, stating that I had a message from the clinic.

"One embryo was deemed viable."

I was worried.

I was worried that something would happen when the embryo was frozen. I was worried that something would happen when the embryo was thawed. I was worried that the embryo would take one look at my uterus and decide it wanted nothing to do with me. Each step of the process brought new things to fret over.

The embryo transfer resulted in a pregnancy, and I rubbed my hand over my belly each day, reassured by many, including my therapist, that everything would be fine. I watched the pregnancy hormones grow, doubling within a normal time range. *Normal . . . that sounds nice.* Nothing had felt normal up to this point, but perhaps my luck was turning. After some days went by, the office called to schedule a viability scan. *Viability scan?*

Interesting choice of words. I called my therapist, and we worked on the anxiety induced by that phrase. Once more, he assured me that everything would be okay.

I waited in the doctor's office wearing a dinosaur-print skirt; skirts always made the scans easier. I looked down and noticed that my tanned legs had faded. They had told me to avoid getting "too hot," so I had spent weeks taking lukewarm showers and avoiding the sun. A physician walked in, someone I had never met. Maybe I had, but it was hard to tell with the mask. *Everything will be fine.*

She inserted the probe and turned the monitor so that I could see. "Nice lining, good thickness." I smiled, as if I had anything to do with it. Then silence. The probe was shifting and her eyes looked like her soul had departed, but nothing could prepare me for her saying, "I can't find a heartbeat." She kept looking, and I propped myself up on my elbows like a backseat driver. There was no point. Any potential baby names were replaced with three words: *incompatible with life.*

My phone notified me of a video visit with my therapist. As I went to the parking lot to take the call in my car, the years of therapy and pills and injections and time off from work and blood draws and hope all crashed down on me. I felt validated, as if I was right to be worried this entire time, and how dare anyone try to tell me otherwise? The call connected, and I screamed.

Skylar Bayer

9

MY BROWN WATERPROOF BOOTS

SKYLAR BAYER

Content warning: heart health

I am sitting in an anchored boat surrounded by ocean, alone. I'm wrapped in a bright red jacket with a neon yellow hood. Supposedly, if I fell in, you'd see my bright tiny yellow head bobbing on the thousands of square miles of dark blue. The boat is open and filled with scuba tanks, an oxygen kit, life jackets, a five-gallon bucket of rope, coolers for samples, and a red and white dive flag, desperately flapping in the wind. Thirty feet away, under water, are my labmates. They are breathing compressed air, wrapped in black neoprene, swimming in forty-degree water—still warmer than the air up here. It's November.

I'm hunched over with my hood up, feeling the ocean swell, hearing waves lap at the hull, staring at my brown waterproof boots. My toes are cold and I feel the deep pull of not being good enough, strong enough, worthy enough. I look at the rising bubbles from my coworkers, indicating that they are still alive. I realize I'm staring at a glass floor instead of a glass ceiling, aching to dive in below and join my team.

I'm stuck in this boat because after six months of scientific dive training I became extraordinarily sick. I was hospitalized, and the doctors learned, after some awful testing, that I have a heart rhythm problem that they couldn't fix. So they inserted a device between my skin and chest wall that defibrillates me when my heart beats too fast. The American Academy of Underwater Sciences, the final word in scientific diving certifications, does not allow you to be a scientific diver if you have a device like this one or a heart condition like mine. When I learned my diagnosis, which was determined from an exploratory surgery, I wept for hours. I felt that my dream of being a scientific diver had been cut out of me with a serrated, rusty edge, leaving a gaping wound in my ego.

A committee member told me that this was my chance to design and lead research projects. After all, an important part of PhD training is learning how to manage a lab. So I helped plan the dives, prepared the divers, focused on safety, and learned how to drive the boat. Surface support is an important job during an emergency; surface workers call the coast guard and administer oxygen to divers suffering from the bends, a heart attack, or a rare shark attack.

My advisor and I once got into an argument about a research proposal. He became defensive and said, "Do you know how lucky you are that you have labmates that will dive for you? I don't think you're grateful or thankful enough." We worked in a lab where almost all of our field projects had some diving, and mine was partly designed before I was diagnosed. I wanted more than anything to dive for my labmates, but I couldn't. I wasn't allowed. Of course I was grateful, and I *did* thank my labmates, and I could give them surface support, but this was a *diving* lab. Diving was where the work was, where the *glory* was.

Back in the boat, serving as surface support for someone else's project and feeling sorry for myself, I hear breaths on the surface. I look up and see three slick, doglike faces and shiny black eyes staring at me. Gray seals. We stare at each other for minutes, curiosity hanging in the air. Suddenly, silently, they disappear. I look around, specifically for a lobster boat. Seals are excellent at taking lobsters out of lobster traps . . . and some fishermen don't have reservations about shooting seals on sight. So seals know exactly what a lobster boat sounds like.

About a quarter mile away, turning the corner around an island, is a lobster boat. It putters toward me, which isn't surprising given the limited number of navigable routes through this particular stretch of treacherous rocks. When the boat is close enough for me to see two men, I wave. I want them to see my flag and avoid our divers below. They wave back and keep coming toward me. As they get closer, I am nervous and excited. If I were being approached by two random men on land, I would be very, very nervous. But out on the water, even though you're much more vulnerable, people are usually just coming by to check on you.

The men are wearing baseball caps, sweatshirts, and Grundéns—the uniform of Maine's fishermen. I can hear their rock music, and I notice they're not wearing their waterproof jackets; they're probably warm from hauling traps all day.

"Hi," they say.

"Hey," I reply with a wave.

"Do you want some hot water? It's not clean or anything, but it's warm. If you give us a bucket, we can fill it up."

"Sure, that'd be great," I say. I'm a little confused at first, but then it occurs to me that it's probably for the divers to dip their cold hands and feet into when they surface. I pass a bucket over the boat railings to them. One guy dips it in the hot water tank and carefully passes it back to me. As I grab the handle and slowly put it on the bottom of the boat, I can smell hot salty water mixed with bleach and see tiny dead crustaceans floating at the surface.

I smile big and say, "Thanks!" I point enthusiastically to the bubbles at the surface. "They will appreciate it when they come out of the water!"

Without pausing, the guy who gave me the bucket casually says, "Oh, it's not for them; it's for you." He's already putting the engine in gear, heading toward other traps, and waving goodbye, all in one motion.

"Oh," I say, "Thanks!" and wave once as they leave.

I'm still in that moment. The warm water wasn't for the divers; it was for *me*. Me, the surface support sitting cold, alone, and useless in a dive boat. Those guys knew that the worst part of a day was waiting to work, to do something, to be useful. I wrestle back tears and survey the beautiful rocky coastline, breathe in the crisp air, and listen to the rhythmic cackles and cries of gulls and terns. Maybe I belong out here after all. I sit next to the bucket of warm water, slowly dip one brown waterproof boot into the bucket, then the other, and start to feel a little bit warmer everywhere.

Furaha Asani

10

HOPE IS MY ANCHOR

FURAHA ASANI

Content warning: phobias, medical procedure, suicidal ideation, grief/parent loss

From my earliest memories I've been aware of something lurking in my mind—an amorphous shadow in the corner of a dark room—that I would later recognise as mental illness. It has taken many shapes, and most days it has been an intense and often debilitating anxiety: I go about daily life on automatic while often feeling that I'm dying on the inside. And I have phobias, many, many phobias—some of which have lasted for a few days or weeks and others for decades.

Since childhood, I've had anxieties about my health. I spent a few years of my childhood scrubbing at my palms and fingernails under a running tap. There was also the time that a small patch of discolouration on my arm had me screaming in fear that I had a fungal infection. These were all habits and fears I outgrew—or so I thought.

My PhD began in October 2013, and my project was ironically based within a teaching hospital, because I'd spent my whole life uncomfortable in medical spaces. While I'd been raised by parents in the medical and dental fields, from childhood I'd always closed my eyes to any reminders of life's fragility.

My hypochondria reemerged intensely just a few months into my PhD, triggered by the sudden death of a friend in my age group due to an asthma attack. My fear started manifesting as a phobia that my throat would spontaneously close. I would wake up at night clutching at my throat. Although my anxiety had always loomed large in the background, now I was shaken by fears of losing my own life while also deeply grieving the loss of my friend.

In a bid to nip this hypochondria in the bud (or stem, because it already seemed to be blooming), I did something drastic.

It was common practice in my department to regularly donate blood and other easily accessible tissues to one another for research purposes. So when my colleague needed alveolar macrophages (white blood cells from the lung), I volunteered immediately for a bronchoscopy. My reasoning was that many people who are scared of heights go bungee jumping. Having health care professionals pass a tube through my nose (after numbing the inside of both nostrils), down my throat (after spraying loads of xylocaine), and into my lung—and being awake throughout, seeing this entire process on a screen—would force me to develop some control over my body and especially my throat.

It was an unforgettable experience. I saw the inside of my lung and my voice box (which looks like a vagina). At two points during the process, when my heart rate went up, I decided to focus my breathing, calm myself, and bring my heart rate back to normal. And that's just what I did. My colleague got the cells for the experiment, and I walked out of the hospital feeling elated and empowered after seeing how my throat functioned in real time.

I did well for a couple of weeks, and then my father unexpectedly died in March 2016 and my mental state worsened. Grief plunged deep into my soul while anxiety wrought havoc in my mind. This time I had the additional fear that, should I die, my death would devastate my already grieving sisters and mother. To my disappointment, my drastic physical intervention, the voluntary bronchoscopy, had no long-term impact on my hypochondria.

Soon anything from a mild headache to an itchy nose was cause for elevated concern and a WebMD spiral. Even my post-doctoral research on a clinical trial developing a new asthma drug did little to alleviate my fears.

The year 2016 passed by in much of a blur. Though my résumé seemed to show that 2016 was one of my most successful years, I don't remember the details of everything that happened. I was functioning on automatic until a panic attack forced me to finally seek help: I was at my part-time tutoring job in front of a room of at least forty students and had just gotten through my lecture delivery when the panic hit me. This "lucky" timing meant that my cotutor took over the class, with no idea what I was going through, while I slunk into the closest chair, trying to control my breathing and quiet the terrible feeling that I was having either a heart attack or a stroke. Immediately afterwards I confided in my cotutor team, and shortly thereafter I took some time off from the job.

In the coming months I received a diagnosis of Generalized Anxiety Disorder and Obsessive-Compulsive Disorder. This was followed by online therapy and an explicit request to my doctors to place me on medication—which was very helpful. One week after starting my medication, I think it took effect. I could hardly believe that this new feeling of hope—and of having the edge of anxiety taken off—was a natural state of being for many people who don't have mental health struggles. My anxiety was at least on a leash.

The natural ebb and flow of my anxiety mean that some periods are much easier than others, and I can go for months without worrying about my health. Periods of intense stress also mean that there is likely to be a barrage of terrifying thoughts, and I am once again fighting to keep my head above choppy waters, refusing to surrender to the deep blue depths of anxiety.

Because of a threat of deportation from the UK Home Office in 2019, I was no longer allowed to be registered with the UK's National Health Service (NHS). Over the years since my fight against deportation began, I became more fearful of falling ill than I ever was when I had access to health care. My anxiety seemed to have seized on the fact that I had no safety net. Once again, this period was accompanied by the fear that I would get sick and die, leaving my family heartbroken. What further compounded and confounded this situation was my occasional suicidal ideation on tough days. On some days I was battling a desire to die while at the same time being afraid of dying: it was a confusing and horrible paradox. During this time my hope had to adapt. It became stubborn. It dug its heels in and harnessed energy from my anger and bitterness, turning them into the momentum I needed to get to the next day, when it promised me that things would be better.

My family and friends are aware of my hypochondria, and I suspect that some people who've worked closely with me have picked up on it, too. It's not something I particularly try to hide. It's also not something I speak about at any opportunity. It just is what it is: a symptom of my deeper anxieties, and perhaps the manifestation of certain aspects of my grief mixed with my reckonings over life and death and how the two overlap.

What has kept me alive this long goes beyond wanting to partake in the lives of my loved ones. Yes, of course I also take pleasure in the big, small, and intermediate—daydreaming about being a plant parent, having whispers of romance, listening to audiobooks, winning auctions on eBay, seeing my friends succeed, and enjoying the evolution of my own career. All of these are seeds of hope. But outside these positive distractions, and when I'm alone in my mind—the same theatre where my most terrifying fears play out—I'm also learning how to intentionally stage-manage a production that keeps me grounded in the moment until all the ugliness passes. So I find myself digging my heels in once again and praying that the seeds of hope I've sown with intention at different points in my life begin to sprout and remind me that there will be better days ahead.

Perhaps one of the most important sparks of hope has come from the self-acceptance that my mind just functions the way it does, and I choose to hold space for myself and my imperfections.

Hope is my anchor. I actively search for sparks of hope and collect them. I find these sparks in many different places—community, faith and prayer, activism, family. I gather these sparks ravenously and hold on to them through the storms.

III

RALLYING THE CREW

When Gabi went to sea, she learned the saying "Ship, Shipmates,
Self." It's supposed to help you remember your priorities while
you are at sea:

> *You need to put the ship before yourself and others because it*
> *keeps you all safe.*
>
> *Your shipmates come before your own desires, because team-*
> *work is key.*
>
> *And you have to take care of yourself to be successful on your*
> *mission or project.*

A s disabled or chronically ill researchers, we often end
up feeling guilt or shame when we have to address
our own needs first instead of prioritizing research—
the ship—or taking care of our shipmates over all other aspects
of our lives. But having a community on board to provide mutual
support lightens that burden of guilt or shame.

To start this section, Alma shares a series of vignettes from
scientific conferences and her field experiences to show that
how people engage and support her as a deaf person clearly

differs in different settings. Alexander suffers injuries from his time in the military, including serious post-traumatic stress disorder that severely impacts his mental health, but his friends in college are seriously concerned about him and he finds help in unexpected places. As a science writer with rheumatoid arthritis, Sophie Fern dreams of being part of a remote field team on an island off the coast of New Zealand. Her dream comes true, and her team want to make sure she can participate in all their adventures, including her first step onto shore. One day, Sophie Okolo, an avid athlete, feels faint after a game of tennis and discovers she has a rare medical condition, and her friends and family start identifying when she doesn't feel well. Closing out this section, Richard, a senior entomologist with a mobility disability, is invited on a field research trip of a lifetime to an Ecuadorian cloud forest and reflects on the importance of teamwork in science.

These stories highlight the importance of literal and metaphorical shipmates, crew, and captains; they involve helpful support from others. We hope this section is empowering for both healthy, nondisabled readers who want to be able to help and for fellow disabled scientists looking for advocacy methods.

Alma C. Schrage

11

THE PLACE I REST

ALMA C. SCHRAGE

Content warning: n/a

ACADEMIA I: CHICAGO, AUGUST 2013

As the conference day inches on, the back of my neck winds tighter with tension and I lean forward slightly, eyes fixed on the speaker.

I am attending my first conference as a junior in undergrad; it's in my home city over the summer. I am too scared to ask for accommodations, so I tell my advisor I will be fine. I attend every session for three days because no one tells me that I don't have to. I sit in the front row at every presentation, eyes darting rapidly between the speaker and the PowerPoint slides. In a quiet room close to a single speaker, I can recognize maybe 4 to 12 percent of the words with my hearing aids; another 30 to 40 percent I can perhaps lipread, and the remaining words I guess based on context. Add additional speakers, background noise, and bad lighting, and I understand nothing.

By the end of each day, I am so physically and mentally exhausted from trying to follow the conversations that I hide in a bathroom stall and cry. For three days after the conference, I cannot read or look at a screen because of eyestrain from lipreading. I am so drained by the experience that I have no questions, no thoughts, no wonder, just a grey exhaustion that covers everything.

FIELD I: PLUMAS NATIONAL FOREST, CALIFORNIA, JUNE 2016

It's 3:30 A.M. I climb into the driver's seat and we're off into the dark on a road that climbs up into the mountains. The crew leader and I have already checked and double-checked that we have everything: GPS, radio, batteries, and all the gear for the bee surveys we must do after the morning bird work is done. I pack an extra item: a pocket-sized WAV format recorder with omnidirectional microphones.

Around 4:30 A.M. we turn off the dirt road onto a rougher one. I maneuver the truck gingerly over foot-deep ruts and boulders as they appear before the headlights. It's close to 5:00 A.M. when I pull over and park. The crew leader and I hurry through the dense chaparral to hit the first bird point count location. Our survey protocol dictates that we must be at the first point half an hour before sunrise.

We get there as the sky starts to lighten. We both pause to catch our breath and let the birds forget the sounds of our passage. I step over a few fallen logs and wedge my recorder into an upright snag, microphones angled up and outward to catch the birdsong in the trees. I hit "record" and then step back to where the crew leader is and take out my clipboard of datasheets. I begin the vegetation assessment, filling out the sheet on my clipboard with estimates of living tree and snag cover as I look around and check distances through my rangefinder. Out of the corner of my eye I see the crew leader noting bird species on his clipboard as he identifies individual birds calling and singing.

A minute later I am done with the vegetation survey; I pull out my smartphone and listen too, but through the oscillating lines of a spectrogram running on my phone screen, where the arc and trill of bird voices look like writing. I note the slightly buzzy trill of the orange-crowned warbler; the slower, more level trill of a Wilson's warbler; and the bold, low, brushstroke song of a black-headed grosbeak or American robin, which I am still trying to figure out how to distinguish. I see other songs: the buzzy chicken-scratch notes of a Cassin's finch and an uncommon treat, a canyon wren, a long sequence of clear notes that look like tilde signs. I see another I don't recognize; I snap a screenshot for reference. A minute later, I see two brief notes that look strangely precise and inorganic, and I know time's up: they're the beeps of the crew leader's stopwatch.

I hop over the logs and grab my recorder, hitting the stop button. I tuck it into my pocket as I step back and quietly list the species that I recognized. The crew leader nods and adds two songs I didn't identify. I think back to one I didn't recognize and translate the visual characteristics I saw, describing them with words that hearing people recognize, such as "clear," "buzzy," "single note," "multiple notes," and "rising trill." The leader nods in recognition, assigning the song to one of the species I missed. The other song might have been too far for my phone to pick up; I may see it at the next survey point.

Through the morning we slowly work our way up the drainage; at each survey point, I repeat the same process: complete the vegetation survey, listen, then quietly confer. In this way I incorporate the new songs into my visual memory so that I can spot the species at the next survey point. The process is not perfect. My phone does not have the same sensitivity as some human ears do, and I cannot pick out the songs the crew leader hears beyond eighty meters. Much later, when I describe these failings to another biologist who sometimes does point counts, she reminds me that even most hearing people are not very good at point counting and tells me to be less self-critical.

The sun is high, and it is beginning to get warm as we finish our last survey point and begin hiking back to the truck to grab our bee survey equipment. Again, the crew leader and I confer briefly: I will hike down the ravine and visit the bee survey points along the upper drainage; he'll drive to the bottom of the drainage and work up from there.

Most people think of my bird listening as a one-hit wonder. They marvel, but they think in absolutes: if I get an identification wrong, I get a pitying look, and if I get it right, I am exceptional. There's no room to make mistakes, to learn. It's rare for other birders to engage with what I am doing. But the crew leader does; he doesn't spend time marveling at what I am doing, the strangeness. He installs the same spectrogram app on his phone, and we argue and compare notes. And every time it's my turn to get up before dawn to help with point counts, I learn more songs.

The truck rumbles off and I am alone in a cloud of dust. I belt my backpack and hitch the cooler strap on my shoulder so that it's a little more comfortable. There are hours to go before I am done, bees to catch and plants to identify, but I am alert as if I have had a full night of sleep. Songs of the unseen world still dance in my mind and somewhere in the branches overhead as I begin to hike down the ravine—using my bee net handle like a hiking pole on the steep slope—toward the stream below.

ACADEMIA II: ESTES PARK, COLORADO, SEPTEMBER 2014

For my last conference as an undergraduate, a little angrier and wiser, I ask my mentor for accommodations, the same ones I get for school—CART captioning. Although it's illegal for conferences to not provide these accommodations in the United States, the mentor is told they're too expensive; when she relays this news, I smile and say I'll manage because I don't want to let her down.

I give my presentation and then attend the only session that is closely related to my undergraduate research. During a networking event, my advisor waves me over and introduces me to an older researcher, who asks me what my plans are after graduating. I reply honestly that I want to get fieldwork experience because I don't have any. He laughs and tells me that fieldwork is not a career. I smile and don't say anything. How do I explain to him that I don't see a career here? He starts talking to someone else, and I drift through the room, watching but with no way to understand and enter the noisy conversations.

I skip the rest of the conference to explore the surrounding mountains. I hike from lake to lake, each one a little higher than the last, until finally I am above the tree line; a cold breeze ripples the glassy water as it reflects the walls of a cirque and a grey sky. A dipper—round and slate-feathered—hops from rock to rock, hunting for invertebrates where the snowmelt flows into the lake. I sit a while with the water, the dipper, and the mountain before I walk on.

Tomorrow I will fly home, and this time my eyes do not hurt and my shoulders do not ache. There are still no answers as to whether I have a career here, but the grey exhaustion is not there and I feel alive with questions and ideas. And that is enough for now.

FIELD II: ELDORADO NATIONAL FOREST, CALIFORNIA, JULY 2018

A blessing of bee research is that bumblebees generally do not give a damn if you are bumbling about making noises as long as you are not in their immediate vicinity.

Field biologists working in bear country learn one cardinal rule: make noise. It warns the bears of your presence and gives them time to move away. Prey animals are not noisy; they move quietly to escape detection. Bears and other strong or dangerous animals do the opposite; they make noise. So the way you move through the world conveys what you are. Be quiet and move cautiously, and you are prey. Be loud, and you are an animal that does not need to hide to survive.

Growing up in a hearing world, sound was never mine. I had to think about the noises I made, constantly being aware of this unseen thing. For a long time, my voice was trained to have just the right amount of volume, tone, and enunciation. To sound deaf—even if others understood me—was something to hide. Speaking entailed constant awareness: be smaller, be quieter, be palatable to other people; do not frighten them with your strangeness.

So I love being in bear country. I become myself without thinking, noisy and free. When I move, the world talks back; logs creak, twigs snap, leaves crackle. The whitethorn rustles and scrapes as I push through it. Some of the long thorns reach my skin despite heavy cotton and canvas, but I ignore them. I move through the world like a bear.

My hearing coworkers have a harder time. Because they hear, they walk quietly without thinking, and all of them have a moment during the summer when a bear comes a little too close, seems a little too interested, and they have to wave and shout as they nervously wait for the bear to leave.

And all summer I crash across the mountainsides; the bears leave me alone and the bees don't give a damn.

ACADEMIA III: TORONTO, ONTARIO, OCTOBER 2019

A week before my second conference of grad school, the hosting university realizes that I am a student visiting from another institution, and it immediately retracts its previous offer of providing interpreters. Like previous organizers, the conference organizer does not have dedicated accessibility funding, so the conference cannot afford to pay for CART captioners or ASL interpreters.

Unlike most conference organizers, this one doesn't offer profuse apologies and tepid assurances; she keeps communicating. In the week before the conference, she cobbles together a team of undergrad and grad students hired as notetakers to type out everything for every seminar and group session; she and the other organizer also install an experimental transcription app on their phones to use as a mic for the presenters.

But it's hard; it's impossible to keep up, and they get tired; their hands get cramps from typing. At one point I see that the organizer, a leading scientist in our field, is typing up the notes herself to give one of the notetakers a break. Her action means a lot; my advisors and hearing mentors always treated accommodations as something they could not be bothered with beyond sending a couple of emails or turning on captions. When these failed, they shrugged and gave up, leaving me to struggle on my own.

The setup works, more or less. The software doesn't work most of the time, but occasionally it works long enough that the notetakers can take a break. The notetakers try to type up everything verbatim; they are students from the organizer's lab and familiar with the content, so what they type seems accurate.

Although this process is far from seamless, it's the first time I've been to a conference where I saw colleagues pulling out all the stops to accommodate me, rather than the other way around. It leaves me hopeful.

Yet I'm frustrated by the complacent attitude of some of the attendees; they offer only sympathy and make no attempt, however small, to change the status quo. Others stare at me as if I am a strange animal and ignore requests to use the transcription app when they speak. It doesn't escape my notice that the two organizers, the notetakers doing all the work, and the people who do use the transcription are mostly women, BIPOC, queer, or some combination of the three; all are people who hunger for change.

FIELD III: ELDORADO NATIONAL FOREST, CALIFORNIA, MAY 2018

It's the end of a solitary month recording birds in the desert; soon I will be in a fieldhouse crowded with five other biologists for a week of training in survey protocols before we can start surveying in earnest. It will be good but stressful this first week before things calm down, with lots of conversations and voices I must try to follow.

I watch the moon set over Mono Lake's tufas at dawn; I enjoy the changing landscape of light across the mountains as I drive north. I try to savor the last of my solitude.

At dusk, I am nearly at the fieldhouse, but I stop just over the mountain pass, parking just off a forest service road. Spring comes in stops and starts here; snow starts falling and I hurry to set up my tent as the ground rapidly turns white. Soon my car, my tent, and the surrounding Douglas firs and red firs are covered. Sound is really just vibration, and when I pause, I can hear with my body the snow's soft silence.

I zip my tent door shut, tuck my glasses and my hearing aids into my hiking boots in one corner, and burrow into the sleeping bag. As I grow warmer, I watch the stretching and shrinking shadows of the trees against the tent walls. In the silence, each cycle of stretching and shrinking tree shadows announces car headlights passing in the night.

I am looking forward to tomorrow.

A slow osmosis seems to happen when I live and work in a small group of six people or fewer, particularly in remote places. This is one of the best things about remote fieldwork. In a matter of weeks, my coworkers gradually adjust to my deaf tempo; sometimes grudgingly, sometimes unconsciously, sometimes intentionally, they become aware of communicating with someone whose perception of the environment is different from theirs.

Why does this happen? Kindness? Of course, but fieldwork is dangerous and difficult, and communication is vital. In a group so small, no person can be seen as unnecessary, expendable; our mutual interdependence cannot be denied. And so they adopt ways of communicating.

They look at me when they speak. They understand that I am not following all the words all the time and cue me in to whatever is being discussed, repeating when I ask them to without condescension, anxiety, or rancor—as if it's a natural thing, which it is. They remember to hold a light above their faces in the dark, to not stand with the sun at their back. They repeat jokes and never say "never mind." They wait until my eyes are on them before they continue talking. They figure out gestures to communicate over distances or follow my directions about communicating on a walkie-talkie (hint: it's twenty questions—beep once for yes, twice for no, and three times if you need me to come over, *stat!*).

Our field crew will develop inside jokes, moments of shared amazement, and many bruises. And I am there in the middle of it—laughing, wondering, complaining—because that slow osmosis makes it a place where I belong.

I feel joy at what I gained in scientific fieldwork, but also anger at how much work there is left in science, academia, and society at large. Fieldwork is my place in the world where I can rest; I'm allowed by the people around me to be both a deaf person and a scientist without denying either. Here the tension at the back of my neck loosens, and I don't have to always think ahead, trying to gauge and accommodate the distance between my world and theirs.

Alexander G. Steele

12

SOMETIMES IT DOESN'T GET BETTER, BUT THAT'S OK, TOO

ALEXANDER G. STEELE

Content warning: suicide, military service, combat

I f I am to believe the captions on my baby photos, I was born into a loving family that really wanted me. I don't remember those photos or the woman holding me, my mother, because she left when I was a baby. I was told that she had a break from reality and thought God was talking to her. I have always had difficulty reconciling those photo captions with my life because the life I've known has been hard. I was passed around and physically abused by my family; no one wanted me—not my father, not my grandmother, no one.

I wish I could say that I chose the military for valiant and noble reasons. The truth is that the military was a way to escape the circumstances of my birth. College didn't seem like a good fit when I graduated from high school. I was bullied for being poor, so why put myself through more years of that?

Out of all my choices for military service, the marines seemed like the best fit. They promised it wasn't going to be easy, but my life was never easy, and difficulty was strangely comforting because it was familiar. The marines was supposed to be a lifetime commitment. After I enlisted, the transition from civilian to full-fledged marine was slow. A very structured pipeline taught me how to be a marine, how to use a weapon, and finally how to do my job.

The human capacity for innovation is a double-edged sword. We can visit the moon, but we've also found new and gruesome ways of killing one another. If anything has taught me that, it was my second deployment to Iraq. The use of improvised explosive devices (IEDs) drove that point home, and I learned that lesson in blood.

I was taught from the movies that when you are dying, you see your life flash before your eyes. When it happened to me, everything was dark and cold, more as if someone had hit the "pause" button; when I blinked, I was no longer on a battlefield in Iraq but in a hospital in Germany. A piece of metal no bigger than a hangnail had caught my left shoulder and I had almost bled to death.

When I returned to the United States a few months later, I was told I was getting discharged. I had suffered a traumatic brain injury (TBI), a minor stroke, and could barely walk. The pipeline from being in service to being a civilian was practically nonexistent, so just as quickly as I returned I was jobless and homeless, and the sum of my possessions fit neatly into a few sea bags and a backpack. Still, I felt luckier than the friends I had buried or those who had lost limbs. I had ten fingers and ten toes, and while I couldn't walk easily, I was physically whole.

That's probably why filing for veterans' benefits never occurred to me. I was never informed that I was entitled to lifelong monthly financial compensation and medical care. I didn't investigate probably because, if I tried to get any compensation, then the problems I was living with—memory issues, chronic pain, not being able to walk right, even post-traumatic stress disorder or PTSD—all those things would go from living solely in my head to being on a piece of paper and then they would be real. After all, I was fine. I'm told that's a coping mechanism.

Instead of trying to figure out how to exist in my new reality, I found work and threw myself into it. I worked twelve-hour shifts for weeks at a time. I'm told that's a coping mechanism, too. My existence was solitary by choice. Who could I talk to about my military life? I had no friends, no family, no one who understood me. No one had suffered the way I had suffered or seen the hell that had been seared into my brain so thoroughly I couldn't even escape it in my dreams. I was alone, and I was okay with that. After all, I had been alone in my earliest memories.

I lived like this for about a year, but I missed being needed. I missed my military family and felt unworthy of my title. Was I still a marine if I couldn't fight? I eventually answered that question by finding a new purpose. If I couldn't get back into the fight, I was going to help others who were coming home broken by developing prosthetics. With a goal like that, I could be around people again and not be ashamed of having the audacity to live instead of following my brothers to wherever death took them. I tried to convince myself that this was why I was whole when so many weren't.

Two years into the mechanical engineering program at my local university, I had a 3.8 GPA, and one term I even took twenty-three credits' worth of classes. I once again threw myself into work, my favorite coping mechanism. Living with PTSD is like slowly bleeding to death without realizing it. I had no idea what PTSD was or why I would wake up sweating and screaming at night, why I kept having flashbacks of being in combat, why I was always on guard for something bad. It had been almost four years since I had been hurt. My routine was fine, I felt fine, and I lived my life because everything was perfectly fine!

And then one day things changed.

Nothing about that day was any different from any other day before it. I just woke up and felt different, as if an invisible weight was on my shoulders that I now noticed for the first time. I felt I had just run a marathon: I couldn't breathe, my heart was racing, and nothing I did seemed to fix the problem. Instead of showering, eating, or going to class, I moved from the bed to the couch and just sat. My body felt like lead. Now I know that I was having one of the worst panic attacks of my life, but at the time I was unaware of why this was happening. I had no idea about the reality of living with PTSD.

I've learned that we all have an internal clock. Every year at about the same time I hit a low because it was my trauma anniversary. The constant work and packed schedule only held it back for so long, and then my brain decided it was time to stop running and start feeling. If it had been for one day, I would've brushed it off and lived in ignorance until the next time. Unfortunately, the night made it worse, and before I knew it morning had come. For almost a week all I did was sit on the couch and try to force myself back to being fine. I begged aimlessly for a reprieve from whatever was happening to me. Even when I did find sleep, it was brief, and I was suddenly in a kind of hell.

Even after a week of living like that, I never thought to ask for help. Mostly because I never had anyone I could turn to for help, why would anyone care about problems that were solely mine? Thankfully, help finally found me in the form of some classmates. One of them had texted me to see why I hadn't been in class, and when I didn't text back a few of them came to check on me. I didn't understand why anyone would care about me, so it felt odd, and up to that point I had thought they were just being friendly with me, not actually my "friends." I was so used to being alone that it never occurred to me that I could even have friends. When they suggested I see a doctor, I learned that I had insurance through the school. Again, it had never occurred to me to investigate what benefits the school offered its students. I saw a doctor the same day.

When I explained that I had spent a week with barely any sleep, the doctor prescribed Ambien. He suggested that all I needed was a good rest and I would be fine. I desperately wanted things to be fine again, so that night I took one pill. But the promised sleep never came. Instead of sleeping, I was in a waking dream, half asleep yet half awake. The next day I couldn't clearly remember what I had done. I had vague memories of what I did, but I couldn't tell if they were real or if I imagined them. That was when I decided that I was tired, not because of the sleep deprivation, but because I couldn't keep existing like this.

That night, I took the entire bottle of Ambien. I don't remember if I called my friends to say goodbye or if a friend came to see how I was doing. I don't know how long I was lying in the tub, nor do I remember much of the aftermath. I didn't ask for the details, and my friends were polite enough not to give them to me.

The transition from trying to kill myself to trying to live was more difficult than the transition from the marines to civilian life. There was a lot of crying, an inappropriate amount of drinking, and a lot of fighting with the government for benefits that I still felt I didn't deserve but that I desperately needed so that I could finally get help.

I remained enrolled in school because I felt that I had to and it was my only source of income. I went from a full-time student to part-time so I could focus on myself and get help. Because of the change in my student status, I couldn't afford to rent an apartment or even a room, so for about a year I lived in my car. During that time my grades dropped drastically. For two years, I was somehow allowed to enroll, and fail, every class I took. It was an ugly transition. It was almost three years after my suicide attempt when I decided to plan my life instead of just survive it.

I still desperately wanted to help people and had completed all my undergrad requirements to get into the school of engineering, so I set out to finish what I had started. Even though I was admitted to the main college, I needed to apply to the engineering program to finish the degree, and the years of bad grades presented a problem. I met with one of the advisors for the program, who told me flat out that I needed to find another degree because my grades were too bad. Instead of accepting that answer, I made an appointment with the head of the engineering program, not just another academic advisor. Something wouldn't let me give up.

I finally got a meeting with the program head. We met in a small office in the back of the main area. It was minimally furnished but mostly filled by a bookcase that looked too large for a space that small. I told the program head about my military service, my PTSD, and even my suicide attempt. It was scary, but he sat politely and just listened. He didn't ask a single question or ask for proof of what I had told him; he simply said he would give me one shot. If I failed any of his requirements, that would be it.

It had been years since anyone had held me to any sort of standard, and I feared I would fail. Moreover, I was worried that if I did fail, he would think I was ungrateful for his help. I don't like letting people down, and, more importantly, I didn't want to let myself down. It was a lot of pressure, but it was a good kind of pressure that came from someone seeing something more in me than just my circumstances. That meeting changed my life, and *he* changed my life, but for once it was a change for the better. I got the grades I needed, and two years later I had my BS in mechanical engineering. I didn't stop: a year after that, I received my MS in mechanical engineering.

Seven years after that meeting, I packed my things and moved across the country, and now I'm a fourth-year PhD candidate studying neuroengineering. I've built prosthetics, published papers, and made advances in my field. I'm helping people now, just as I imagined when I first decided to start this journey. I even went from living in an old beat-up car to owning my own home in five years. Best of all, after eight years of fighting with the government, I was finally informed that I was totally and permanently disabled because of my service.

Of course, I wasn't as lucky as I had tried to convince myself I was. It felt good to be told I wasn't being overly dramatic. I now get monthly disability, health care, and access to professional help, which I have desperately needed. It's not all good news, though. I live with aphasia, memory problems, test anxiety, stress, depression, PTSD, and many other problems. I am learning to live within my new limits and am adjusting to my new reality.

This sounds like a happy ending. It also sounds like a story of perseverance in the face of overwhelming odds, but I don't feel that this is that kind of story. Trauma leaves scars. I have several scars on my brain, scars on my body, and scars wherever it is that *I* exist in this big sack of meat. None of those scars have gone away because my circumstances have changed. I find a type of joy in my research and education that, for the most part, keeps me balanced on the perilous tightrope of good mental health. Even then, every single day, I wake up and plan, often in vivid detail, how I will kill myself. Then, every single day I decide that it's probably best that I don't kill myself . . . probably.

The feeling of wanting to die follows me around. It's a thought, or maybe a whisper. It's something I've grown quite used to after all these years. I don't know where it came from or why no medication has managed to touch it. That's the thing no one warned me about, and I find that sometimes my outside reality doesn't quite align with my inside one. Things have gotten easier, but they don't "get better," and it's taken me longer than it should have to realize that that's okay, too. It isn't a comfortable existence, but it's mine and I accept it.

I'm telling my story because you may hear a similar voice in the back of your head that wants you dead. Maybe you also feel alone, broken, damaged, and scared. I'm telling you my story because, after all the hardships, anxiety, depression, and suicide attempt, the one thing that hurt the most was the thought that I was the only one, that I was so broken that I would never find the magical place of "better."

The good news is that I know I'm not alone. And now you don't have to be alone either.

Sophie Fern

13

CHRISTMAS ON RANGATIRA ISLAND

SOPHIE FERN

Content warning: n/a

When I applied to study marine biology at university, my careers teacher called me into her office for a chat. She showed me a letter that she'd just received about my application that asked whether she felt that I would be able to physically handle the course, knowing about my condition. I think that the phrase "getting on and off boats" was used to describe a possible scenario in which I might have problems. But getting on and off boats was the reason I wanted to take the course. I've had rheumatoid arthritis since I was fifteen, but I'd loved wild places and adventures since well before that. So I brushed the question aside, saying, "Yes, of course! I'll be fine." The answer, really, was "I'm not sure. Some days I might need help. And some days I won't be able to do it at all." I wasn't brave enough to say this in case this door would be shut in my face.

Many years later, I'm standing on the bow of a tiny fishing boat that's bobbing up and down in the ocean swell off a rocky island about 650 kilometers east of the mainland of New Zealand, and I remember that letter. I'm the last person to leave the boat because I have been delaying the evil moment by pretending to be extraordinarily busy with something important on the stern.

I'm now waiting for the skipper to tell me when to jump. I've been practicing this leap for months and have worked out that I should lead with my dominant right leg, even though my right ankle is now pinned into a permanent walking position with rather a lot of titanium. Although I have less balance on my right side, I'm still more comfortable leading with it. Not that much actually feels comfortable with rocks in front of me and the Pacific Ocean slopping gently around the boat.

When the skipper calls "jump," I do. He has nosed the boat so close to the rocks that the leap that seemed an Olympian feat in my mind is only a large step that is over in a moment. I've survived the first obstacle of remote fieldwork—landing on an island with no wharf. The team on the shore grabs me and leads me up the rocks to safety. No one wants to start the field season by fishing someone out of shark-infested waters. And I didn't want anyone to regret that I was there.

Rangatira Island is a nature reserve and the third-largest island in the Chatham Islands archipelago. It is home to millions of breeding seabirds as well as New Zealand's largest spider. I was on Rangatira that summer to help study the nesting behaviour of a bird that is conservation royalty—the Chatham Island black robin. In 1985 the black robin was the rarest bird in the world, with only five individuals left. Only two of the birds were female, and only one was able to lay fertile eggs. She was nicknamed "Old Blue" for the blue band on her leg, and all of the three hundred or so black robins that are alive today are directly related to her.

This was not my project. The team was led by my friend Mel, who had a grant to bring a writer with her to communicate the work that the team was doing on the island, and I was that writer. Despite dreams of spending my life in wild places, I cannot have a career that involves extensive fieldwork when

I'm not sure which of my joints is going to fail next. Arthritis has involved a gradual progression of pain and swelling and then loss of function that mostly never returns. My wrists have ossified, I'm on the waiting list for a shoulder replacement, and I haven't straightened either of my arms since about 1995. I have trouble walking on uneven ground, and I can't carry anything heavy. But working with people who know me makes everything easier because they see me as a whole person rather than just the warning flag of a diagnosis. I've also built a reputation as a competent field trip cook and excellent communicator, both valuable skills in the field.

Before we started work that first morning, we needed to get equipped with boards. These are thin, letter-sized sheets of wood with ski bindings that we attach to our walking boots to spread our weight more evenly on the ground. Rangatira Island is pockmarked with seabird burrows that make the soil friable and likely to collapse. It was late November when we landed, and many of the burrows were still occupied, mostly with broad-billed prion chicks that stayed hunkered down during the day while their parents were fishing out at sea. Our work took us off the main path deep into the bush looking for both black robin and other bird nests, so we needed to wear these boards for most of the day. We soon developed a wide "cowboy that just got off a horse" stance and a slightly swinging gait to make sure that we didn't trip over our own feet.

There were only four of us on the island that summer, so, even though I was there mostly to write, I also helped with the fieldwork. Each morning Mel divided up the nests to be checked and we filled our packs with equipment, food and water, and, in my case, painkillers and headed out into the bush until lunchtime. I spent my first few days on the island lost and, despite the boards and my best efforts, was often reconstructing the prion burrows that I'd collapsed. I scrabbled through the dirt on my hands and knees until I could grab a chick, which I put into a cloth bag that I hung in a nearby tree. Then I dug out the excess soil and made a solid new roof for the burrow with sticks and leaves. I was often pecked as I gently persuaded the now slightly grumpy chick back into its new burrow. Then I packed everything back up and headed off, sometimes to collapse another burrow two or three steps away.

Our morning job was to monitor the black robin nests to check how the eggs and, later on, the chicks were progressing. My favourite nest was home to a chick that I had nicknamed Pip. It was the easiest nest to check because it was close to the path and, as it was in the stump of a dead Chatham Island *ake ake* tree, was also close to the ground. For once I didn't have to crawl, wriggle, or climb to see into the nest. Pip was the one surviving chick in this particular nest. He was huge, and I imagined him to be like one of those spoilt kids who know that they have a good thing at home and plan never to leave but still have not told their hardworking parents. By the time we left the island, Pip, true to form, was two days overdue to fledge.

In the afternoons, while the others went nest hunting or mist netting, I sat on the back deck of the hut writing something witty and educational about the science or the island to send back to Science Outreach at Canterbury University for the project blog. Every few evenings I took the satellite phone out to the rocks where I'd first landed, made sure that I was facing in the right direction, and kept my fingers crossed as I sent out some of the world's more expensive emails. I often stayed on the rocks well after the emails had finally been sent, watching a fur seal playing in the water or the skuas circling their nest or, one memorable night, two giant petrels pulling apart a dead seal carcass.

On Christmas morning we worked as usual, checking nests. Many of the young were close to fledging, so it was a busy time for the birds and for us. Having the rest of the day off for the holiday, we packed our Christmas lunch of cheese sandwiches and prunes and climbed up to the highest peak on the island. It was a stunning midsummer day, clear and sunny; apart from one farmhouse visible far in the distance on Pitt Island, we could have been the only people in the universe. That climb to the peak was a hard slog for me, however, and even after a dip in the deep rock pools later that afternoon, my ankles were sore and I was exhausted. So I wasn't keen to join the next holiday adventure, a sleep out on the Trig, the second-highest peak on the island. I would slow everyone down, I argued, and they should just go without me. And it was just for fun, so I didn't really need to be there.

The team, however, refused to listen. Provided I could walk, they said, and I obviously still could, we would be having this adventure together. And so a few hours later we hauled ourselves and our stuff up to the Trig. From there we had a perfect view down to Woolshed bush, where the hut was, and behind us was the summit, from which a disapproving brown skua was glaring at us. There was enough flat space on the Trig for the four of us to lay out our mattresses, and we snuggled deep into our sleeping bags. As dusk fell, tītī circled us, unsure what these strange lumps were doing on their previously uninhabited rocks. Then, as darkness really fell, the prions and petrels returned from the sea, flying into trees and tumbling to land before waddling off to their burrows, ignoring us by walking over or into us. I didn't sleep much that night because of the birds and the cold and slightly lumpy ground, and, as dawn broke, I could hear the chorus of forest birds in the bush below us. We were up soon after and packing to walk back down to the hut for breakfast and another full day's work.

Being brave enough to say yes to that extra adventure felt like I was fulfilling my childhood dreams, obsessed with boats and islands and adventures.

I'll be forever grateful to the team for not listening to me that night. My adult instinct can be to say no, to be overly cautious, to not push my luck, and never to ask for help. Just being on the island was a privilege, and I would be missing out if I didn't go with them, but that's life, right? But that night they wanted me on the Trig with them, and, by carrying my mattress and encouraging me up the hill, they made sure that I could be there.

Even now, many years later, when I'm trying to find peace and calm, I think of Rangatira Island. I imagine the robins foraging for insects to feed their chicks and the shorebirds busily patrolling the rocks and defending their territories. I can imagine myself there, sitting by the small stand of flax bushes above the rocks at Whalers Bay. I'm having a quick rest looking out to an empty sea before attaching the boards to my walking boots and heading into the bush to look for nests and birds.

Sophie Okolo

14

LIVING WITH A RARE CONDITION

SOPHIE OKOLO

Content warning: diet, food, weight

O*ne, two, three, my body is falling on the grass. I can feel the afternoon heat, my slow breathing, and my head still spinning as if in a daze.*

It was the second year of college, a period marked by intense personal and professional trials. But intramural tennis was a fun and creative outlet to relieve my ever-increasing stress; I've always loved playing tennis. Tennis was also a relaxing way to get to know other students from different programs.

On the first day of class, I couldn't contain my excitement and talked nonstop with my roommates about improving my tennis skills. But the hot day was going to worsen what was to come. I arrived on the court, met the instructor and other students, and began playing a few drills and games. Once the class ended, I started to feel lightheaded. Since it was a hot day, this didn't concern me at first. When you are pushing too hard during a workout, your blood pressure can drop or you can become dehydrated, which can cause you to feel light-headed, dizzy, or faint.

I thought I wasn't overexerting myself, but what happened next still feels like something in slow motion. The game was over, and the class ended. I made small talk with a few students, and then everyone left while I stumbled across the court. I don't think anyone saw me, and I was trying not to make a scene. After a while I felt dizzy and lightheaded and wondered if it was really the heat. I exited the court to a nearby grassy place and passed out.

I am not sure how long I was passed out, perhaps only some seconds, but I finally regained consciousness. I didn't get up until I rested for a few hours for fear it might happen again, or maybe because I knew my body needed rest. When I came to, I knew it wasn't just the heat and I had to get to the bottom of the issue. I spent hours googling my symptoms to see if there was a condition that matched or if everything I was experiencing was a result of an intense period in college. I also had several discussions with friends and family to try to understand what my symptoms could mean.

At first my family did not understand what was going on. I am the only one in my family to have this condition, so we did not know how to deal with it. They always told me to eat more or mentioned how much I exercised. They also talked about my fluctuating weight every time I visited or when they saw my picture. I knew they meant well, but I felt confused and embarrassed. In my mind, I felt and looked healthy, but my family and friends worried about my low weight and fainting spells. At the end of my sophomore year, my family urged me to make an appointment with a doctor.

Since I rarely visit the doctor, this appointment was quite monumental. I didn't think I needed to see a doctor because I was also in denial about how bad my health was. I didn't want to know what was wrong because a medical diagnosis could lead to more anxiety and stress. I wasn't ready for that. But I also knew that delaying could make an easily treatable condition worse or untreatable.

When I finally went to the doctor's office and had my blood drawn for a glucagon test, the doctor was shocked at my dangerously low blood sugar. The test confirmed that I didn't have diabetes, but I was diagnosed with nondiabetic hypoglycemia, a rare condition of low blood glucose in people without diabetes. I had a severe form of hypoglycemia that could be more harmful than diabetes or hyperglycemia. Untreated hypoglycemia can cause seizures, fainting, and even death because the brain cells stop working without glucose, and the problem requires immediate intervention. I was relieved to have a diagnosis but felt uncertain about how I was going to manage this rare condition.

Although I knew my diagnosis, I was in college studying for a science, technology, engineering, and mathematics (STEM) degree. I didn't have time to properly treat my low blood sugar. My program was filled with labs, all-nighters, and overwhelming stress that only worsened my condition, significantly affecting my physical and mental health. And while the gym and school clubs provided moral support, dining was tricky because there were few healthy options except the salad bar.

Much like when going down the cereal aisle in a grocery store, I was unsure about what to eat and what to avoid, but I needed to eat a consistent diet to maintain my blood sugar. At times I risked getting constipated or having a food allergy because I needed to eat meals. While a complete diet overhaul was crucial, I still had food cravings and wasn't ready for it.

All my friends and family now know that when I get quiet I'm probably having a hypoglycemic episode. They start going through my purse or theirs for glucose tablets, food, or snacks, or they change the route to get to the nearest restaurant. Sometimes they ask if I need anything. Fruits high in sugar increase my blood sugar levels immediately.

In the fall of 2015, I forgot to eat breakfast because my family and I were late for church and rushed out of the door. I didn't remember to carry a snack, which I often try to do. There was a lot of banter in the car until I suddenly got quiet. I knew what this familiar feeling was. I felt limp and as if I was going to pass out. My brother noticed that I looked faint, and my family quickly asked what was wrong. But speaking required energy that I didn't have or couldn't summon. My sister finally said that I was having a hypoglycemic episode. Instead of rushing to church, we drove to the nearest 7-Eleven. They bought some snacks and even helped me open a bottle of juice. As I drank, I started to feel better and more relaxed.

My best friend has experienced a similar though not as serious event and has helped me drink some juice when my blood sugar has dropped. My family had experienced similar moments before, but everyone was scared after this church episode, especially me, because my breathing was shallow and I couldn't speak. This event deeply affected my family and changed the way they viewed my condition and helped me manage it. After that, they encouraged one another to carry snacks everywhere, and we prepare ahead of time for low blood sugar. Still, managing this condition is not straightforward, and I have difficult days. It's a lifestyle change, like dealing with my stutter. I now have a greater appreciation of food and have learned to avoid skipping meals. Even exercise has a new meaning. I always have to listen to my body, even if the message is small and seems insignificant.

In 2020, on the brink of the coronavirus pandemic, I transitioned to a plant-based diet because I wanted to control my blood sugar better and live a healthier lifestyle. A plant-based diet is easier to control because the composition of nutrients is better known and less complex than in the highly processed foods that are common in our culture. Plant-based or plant-forward eating patterns involve fruits and vegetables, nuts, seeds, oils, whole grains, legumes, and beans. Changing my diet to one that I had more control over was a significant change because it was hard to eat fewer delicious foods like orange soda, hot chocolate, ice cream, and tasty desserts. But the result was little to no bloating, gas, and abdominal pain—conditions that had become common because of my hypoglycemia.

While nondiabetic hypoglycemia does not define me, I am aware of its implications, the choices I make, and the lifestyle I need to maintain for healthy aging. Now in my thirties, I'm more conscious of exploring what these choices ultimately look like. If I could go back several years, I wish I had had the mental, emotional, and physical support to manage my condition during my time in college. My family lived in a different state and I had no direct relatives in the area, which made college a lonely experience. Also, my major in bioinformatics was already challenging, and having nondiabetic hypoglycemia doubled the pressure.

Talking about this condition is often difficult because some people assume that I have diabetes when I do not and also because I look "healthy." I'm also a five-foot, eleven-inch, athletic girl who ran track and still loves sports. Others do not see that my handbag has a lot of bars, drinks, glucose tablets, and other items to ward off fainting episodes. I have learned that those with nondiabetic hypoglycemia often suffer in silence or downplay the seriousness of their condition.

I often find it hard to talk about hypoglycemia because some people see it as just having low blood sugar. When I mention my symptoms—headaches, extreme hunger, dizziness, and blurred vision—then they realize how severe it is. I feel defeated when this happens because I seem to be discussing something insignificant until they learn what's going on. I'm not afraid to talk about diseases with family and friends, but ironically I avoid discussions about hypoglycemia.

Depending on the day, life with nondiabetic hypoglycemia is manageable but unpredictable. I am not sure what caused my condition, how long I will have it or continue to have to manage its symptoms, or when I need to see a dietitian or doctor for blood tests. Ultimately, my goal is to continue living a healthy and more sustainable life, especially as I get older. As the proverb states, health is wealth. Indeed, I agree.

Richard Wendell Mankin

15

PLANNING THE JOURNEY OF A LIFETIME

RICHARD WENDELL MANKIN

Content warning: n/a

Would you pack up a trunk of microphones, vibration sensors, amplifiers, recorders, petri dishes, notepads, and a laptop and then fly to a remote part of the Ecuadorian Andes to study fruit fly courtship? Well, if you're doing research to control insect pests by managing their mating behavior, you might jump at a chance to join this expedition.

In this case, the pests were tephritid fruit flies, which are economically important agricultural pests worldwide. Females of many tephritid species lay eggs on developing fruit, and the hatched larvae burrow in and eat or damage the fruit before the grove owners can harvest it. Before the females lay their eggs, they mate with courting males on the undersides of the leaves of a fruit tree. If farmers had economically viable means of interfering with fruit fly courtship, the cycle of adult mating, egg laying, and larval damage to fruit would be disrupted and the currently heavy use of pesticides with potentially harmful human and environmental impacts could be reduced.

Unfortunately, there is no one-size-fits-all solution to disrupting tephritid mating. Each species has a different pattern of courtship, during which males gather into groups on the undersides of leaves, displaying and buzzing their wings in distinctive patterns while interested females fly up, land on the leaves, and pick a favorite. The better we understand these courtship patterns, the better we can develop methods to disrupt them.

Ecuadorian forests are very biodiverse. One cloud forest includes a "hot spot" that contains a cluster of fifty closely related tephritid species in the genus *Blepharoneura* that feed on melon fruits such as watermelon. Unfortunately, much of the cloud forest region is threatened by climate change and human population growth. The stakes are high to examine the mating behaviors of the tephritid species in this diversity "hot spot" before the forest changes forever.

It's not easy to do research in a remote cloud forest in Ecuador, however. First, relationships must be established or strengthened with researchers at Ecuadorian universities and persons familiar with the local geography and wildlife. Then governmental permissions must be obtained to conduct the research in a specific area. Finally, the researchers must plan out and fund the lodgings, equipment, and supplies in advance. I was one of eleven researchers who carried out the project in 2019.

The research is even more difficult when one has limited mobility.

Let me backtrack to the 1950s, when I was living near an air force base in New Mexico. I developed an intense curiosity about the deserts and wooded mountains ten or twelve thousand feet high, as well as their insect inhabitants, nearby farms, jets, radar, and other technology. I remember spending time outside watching airplanes, birds, and insects flying above and staring down at ants and insects on the ground, thinking, "There are patterns here." I wanted to understand how these aspects of technology, nature, and animal behavior fit together. That desire to understand remains with me today, seventy years later.

I grew up before the passage of the Americans with Disabilities Act, and access for those with disabilities was not required at schools. Missing muscles in my legs and arms impaired my mobility, and the school system provided "bedside teachers" who schooled me at home until the end of the second grade, when my teacher convinced my parents and my school principal that I belonged in regular classes. Her helping hand started me on the path through grade school, college, graduate school, and a postdoctoral associateship.

As a child, I was unsure whether limited mobility would affect my ability to succeed as a researcher. Although surgery at age eleven to straighten my legs allowed me to walk with braces and crutches, my mobility remained limited. My legs did not bend while walking, so I moved them out and around on each step, and it took a few years to learn how to travel over rough ground and to climb steps. I had to prepare ahead and set up workarounds to access buildings or deal with longer times for traveling between classes. Along the way, I learned that school and career success depend on careful planning, aided by support from parents and family and later from friends, colleagues, and helpful bystanders. Planning helps me break barriers, especially when visiting a new environment.

By the end of my postdoctorate, I was initially focusing on specialized technologies for studying insect behavior and neurophysiology, and then the focus expanded to insect pest management and environmental conservation. Toward the end of this period, Rachel Carson published her book on the broad and harmful environmental impacts of DDT—the famous pesticide that was used during the World War II era to control insect damage on crops and to protect humans from malaria and other insect-vectored diseases.

I was fascinated with how we could use new technologies to learn about insects in ways that would improve our agricultural and public health practices. After being hired as a research entomologist in 1980 at the USDA Center for Medical, Agricultural and Veterinary Entomology, I became interested in its insect acoustics research laboratory, which focused on detecting hidden insect infestations in stored products, trees, and soil, as well as on disrupting the mating of insects that locate one another using acoustic signals or vibrations carried along leaves, stems, and branches. With colleagues and students, I have developed and tested instrumentation and signal analyses that detect and interpret sounds produced by hidden insects moving (which sounds like scraping) and feeding (which sounds like snapping plant fibers) in distinctive patterns.

Many insects use acoustical or vibrational communications for mating behavior, some of which, like fruit flies, crickets, cicadas, and stink bugs, are familiar to many of us. We collected data on both the sounds that Mediterranean and Caribbean fruit flies made and their temporal patterns and then analyzed their mating communications. We often had to collect signals at remote field sites because the insects could not always be reared in the laboratory or transported to it.

Fieldwork is necessary to ensure that one is not working on the wrong problems, to understand the important processes that help or hinder one's pest management goals, and to make sure that lab experiments work in the real world. I have eagerly collaborated with colleagues in several field studies. And even though I've always had limited mobility, I've exercised regularly and learned to prepare for hazardous terrain at field sites. Depending on the field environment, such as Florida citrus groves, I may have to construct lightweight benches to flatten the working surfaces and keep the equipment dry.

The researcher Marty Condon has been studying the neotropical *Blepharoneura* genus for decades. These flies feed on diverse *Gurania* plants—cucumber relatives. Very little is published about the mating behaviors of *Blepharoneura* species and how they compare with the behaviors of important documented fruit fly pests. Marty enlisted the aid of Sonja Sheffer, a molecular biologist, to genetically identify several species because they could not be distinguished morphologically, and she also asked Sonja to examine the mating behavior of *Blepharoneura*. Sonja called me about the equipment and procedures needed to record and analyze the mating calls, but after some discussion, she recommended that I come along and assist in person.

Soon I was flying to Quito and heading to a lodge near the Sumaco volcano in the cloud forest where Marty, Sonja, and nine other ecologists, biologists, and students on the research team were searching for host plants, raising fruit flies, and collecting DNA for species identifications.

Upon arrival, there was an immediate change of plans because the cloud forest was too wet and noisy to use acoustic and video equipment that was only partly waterproofed. I'm used to holding my crutches vertically on rainy days in Florida to prevent the tips from sliding away, but the shadiest surfaces in the cloud forest were almost like ice—much more slippery than I'm used to. While we were disappointed at first, we quickly decided to conduct the observations in a quiet dorm room in the Sumaco field station lodge, away from most human activity.

Over a three-week period we studied the visual and acoustic mating signals of the different tephritid species that had been found in the forest and whose offspring had been reared on *Gurania* plants in the lodge's kitchen. Individuals of these species were about the size of a typical housefly and fit easily into a petri dish. A little bit of beeswax was used to attach the accelerometer sensor to the dish and pick up the walking and fanning activity of the courting pairs, as well as their low-frequency abdominal vibrations. The daily rains pounding the metal roof and the frequent bird calls made it difficult to collect acoustic recordings of fruit fly matings, but we could detect their vibrations with the accelerometer and observe the visual displays of the male calling from underneath the lid of a petri dish (as if it were the bottom of a leaf). We sat our small accelerometers next to the three-inch petri dishes, quietly waiting and watching for male and female pairs to perform courtship behaviors.

During most courtships, the female eventually declined the male and flew off to the side of the petri dish. However, after six days of observation, we finally observed the first mating. I excitedly called over everyone who happened to be in the lodge, and we all observed and cheered (the flies didn't seem to mind).

I came back to the lab with dozens of vibrational communication recordings, observational notes, and meaningful memories of working with other scientists and students in the cloud forest. Marty, Sonja, and other group members helped me plan the experimental details, set up the apparatus, and maneuver around the difficult terrain outside the lodge. Now I'm analyzing the visual behaviors and auditory signals produced by the different *Blepharoneura* species.

Science is so slowly planned out that only rarely does a person become aware of an important, imminent opportunity, head off to a remote cloud forest with colleagues, and then successfully collect data from just one summer field season. Our success was due to good luck and the collaborative efforts of many individuals. As anthropogenic development encroaches on the habitat of the *Blepharoneura* fruit fly, the data we collected may become a rarity. Looking back, I think this may have been one of the last possible opportunities to capture one of the first and possibly last recordings from all these threatened species with modern equipment. Soon the cloud forest habitat may be changing drastically due to the impacts of climate change and lead us further into unknown territory.

I feel honored to have been a part of this project and its success. The beauty and the diversity of the Andean cloud forest, its birds, and its insects are spectacular. While the forest remains, more people, including other scientists with disabilities, should be able to experience it. We all need opportunities to explore, ponder, and pass on the insights we have gained.

IV

IN THE HEART OF
THE MAELSTROM

*"Maelstrom": a powerful, violent whirlpool or vortex typically
in water; a situation or state of turmoil and agitation.*

D uring some events in our lives, we feel that we're fighting through a powerful whirlpool of conflict, terror, and
turmoil. In the disabled community, these moments of
conflict can be more common because of the inaccessible world
we live in. The built environment, the social attitudes, and the
culture of academia are either actively fighting against us or simply not constructed with disabled people in mind. Even "small"
challenges stack up if we face them frequently; the frustration
can loom like hundred-foot waves that we have to evade or dive
under to survive.

To begin this section, another anonymous author, a PhD
student in biology, humorously shares her story of discovering
that she has pelvic floor dysfunction, something that affects her
work and her relationship with her boyfriend. Juniper is an ecologist and activist in the middle of protests against police brutality
in Portland, Oregon, but has to be separated from their much-
needed service dog, Wallace. Syreeta is starting a new semester,

balancing her classes and health, and then goes to the doctor for a medical test that suddenly goes extraordinarily wrong. Amanda, having suffered a brain injury years ago, is now in the middle of her PhD candidacy exams asking for accommodations that have never been requested before in the department. Stephanie has recently been diagnosed with multiple sclerosis and is attending a large conference in New Orleans in the manner she always has when she suddenly gets lost and her foot starts to drop. Ending the section, Divya had always wanted to be a planetary geologist, but her plans start to crumble when she gets sick during college.

These stories focus on getting through difficult events such as diagnoses of medical conditions and trauma from medical procedures, or on navigating typical scientific activities such as conferences and graduate school exams with a disability.

Deflating balloon

16

THE BUTT BALLOON

ANONYMOUS 2

Content warning: medical procedure, pelvic pain

"Okay—it's time to push."

I want nothing more than to internalize those five words. But dressed in a flimsy, translucent hospital gown with my knees spread wide, I've never felt more ill-equipped for a task.

I open my mouth to tell the nurse I'm not ready, that I'm scared and not sure if I'm strong enough to do this. But it's too late; she's already walking away. She draws a curtain between us, and I hear the beep of a timer as she sits to listen for the telltale plop of rubber hitting the bottom of the plastic toilet beneath my rectum.

It's October 2016, and I'm in a hospital in downtown Boston for an anorectal manometry procedure—a fancy term for a test of my ability to push an inflated balloon out of my ass without phoning a friend. Let's be clear: That is something everyone should be able to do. It's a hell of a problem if you can't.

To come in for this test, I snuck out of the lab where I work as an infectious disease researcher about three blocks away. No one knows I'm here except for my partner Matt, who, just this morning, I insisted not come with me; this was something I'd get through myself. I'm here because, for the past year or so, I've been suffering from a pretty bizarre constellation of symptoms. The inside of my vagina has been feeling like it's constantly being scrubbed with sandpaper. I also need to pee—all the time. And it's become not at all uncommon to go three or four days without a bowel movement.

This combination left most doctors scratching their heads. For months, I got the same set of questions, over and over. "Have you given birth recently?" No. "Have you tried eating more fiber?" Yes. "How many sexually transmitted infections have you had?" None. "Have you just tried using lube?" Have you?

It took seven months and eleven doctors for someone to finally give me a diagnosis: pelvic floor dysfunction. Basically, that means that the muscles around my pelvis—which control pretty much everything down there—spasm at random. In other words, I have no control over my anus, my urethra, or my vagina.

Functionally, that means three things. One, I hold on to my poop 24/7. Two, when I need to pee, I need to pee. And three, because my vagina can't relax, sex with Matt has become a futile exercise in trying to fit a . . . let's call it an average-sized peg into a very, very narrow hole.

But even that diagnosis wasn't definitive. To confirm pelvic floor dysfunction, I needed an anorectal manometry test to prove that I can't defecate at will.

When I first heard that, I laughed. Something that absurd could not be a real medical procedure. But indeed it is, and that's why I'm here in this room with a balloon up my butt. Not three minutes ago, that same nurse watching the time tick down laid me on my side and pushed the balloon—still flaccid—inside of me and pumped air into it with a little syringe. And voilà: an inflatable tampon for my ass.

I have two minutes to "evacuate" the balloon from my body. But sitting here, in nothing but this flimsy gown, my mind is swimming with the sounds of the clock on the wall and the faint smell of bleach on the linoleum floor. And I can't help but feel that this balloon rescue mission is dead on arrival.

But I close my eyes, and I push. I push because I'm sad and angry and humiliated. Because I've gone from being the person in my lab who runs the most experiments in a day to the person who runs to the bathroom the most times in a day. Because I've laid myself down on the floor of that same bathroom I share with my coworkers to give myself enemas during my lunch break.

The last time I took a hard look at my life, many parts were unrecognizable. In the past year, I've had to stop riding my bike and wearing tight pants. Hemorrhoids and bloody underwear have become daily fixtures. My suggested buys on Amazon have morphed from running shoes and short story collections to incontinence pads and squatty potties.

Worst of all, in the past year I've sort of shut down. My symptoms just aren't the kind I can talk about. Bringing them up at work would be a form of sexual harassment, and I'm scared that if I tell my friends I'll come off as whining about a small, insignificant nuisance. The old me wasn't afraid to talk about her vagina to her friends and her partner. But now, even with Matt, I've gone quiet. When we first started dating, Matt would joke that I was the least clingy person he'd ever dated. He'd tell me he admired how independent I was and how I never let anything stand in my way. So how am I supposed to tell him that I'm not sure that independent, self-assured person is still around? Whenever I go looking for her, all I see is the person who needs to stop her partner in the middle of sex or wake him up at two A.M. because she's wet the bed. So I bottle everything up. For me, the person whose every orifice is knotted shut, I guess that's just the easy thing to do.

So sitting here, on this cold, plastic toilet, I push because this feels like my last chance. If I can do this, if I can just do this, maybe it'll all be over and everything can go back to normal.

But nothing happens. The balloon doesn't budge.

Eventually, two minutes pass. The balloon, on the other hand, does not. When the timer goes off, the nurse steps out from behind the curtain and tells me what I already know: I've failed the balloon test, as if I'm some four-year-old at a birthday party who's done a shitty job at inflating a balloon.

But one way or another, that balloon has to come out. I waddle back to the cot in the middle of the room and lie down so the nurse can reach inside me and pull out the balloon that I've failed to expel.

After she's gone, I wipe off the petroleum jelly from between my legs. I put my clothes back on. I walk back up to the reception desk. I hand over the $35 I owe to Beth Israel Deaconess Medical Center for the butt balloon that, for the record, I didn't even get to keep. And I make it all the way outside before collapsing behind a bush and bursting into tears.

There, in the dirt, I lie on my back and cry. I think about all the small, everyday things I always took for granted that I can no longer do and the person I was, or that I thought I was but definitely no longer am. And I fill up with hatred, for the kids whizzing by on the bikes I can no longer ride, for the happy couples in restaurants holding hands, for the three male doctors who asked me to consider that pain during sex may be normal for some women. But most of all, I have hatred for myself, for not being able to just take a shit. Lying there in the dirt, I feel these bits and pieces of myself being chipped away. And I cry until everything inside me has burned out.

I don't know how long I lie there behind that bush. I barely remember standing up, but somehow I do. I pull myself up and tell myself to move because no one else is going to do it for me.

As I walk toward the bus stop, my phone buzzes. Matt has texted me. It's not a response to anything—just three words: "I love you," with proper capitalization and a period at the end, because he's fancy like that. The text knocks the wind out of me. All I know is that I'm tired of feeling alone.

When I get home, the floodgates open. I tell Matt I'm angry. I tell him I'm scared. I tell him there are days I can't tell if I'm losing control of my mind or my body or both. I ramble for what feels like an eternity until the words just run out.

When it's over, Matt doesn't say a thing. He just comes over to me and pulls me close; his embrace is warm and safe and familiar, and he smells like home. I wrap my arms around him. For the first time in a long time, I know I'm holding on to something tightly enough.

Juniper L. Simonis

17

THIS IS WALLACE ALFRED RUSSEL SIMONIS

JUNIPER L. SIMONIS

Content warning: police brutality, trauma, chemical weapons, murders of disabled individuals

In late July 2020, during the height of the racial justice protests in Portland, Oregon—as I lay in bed recovering from being brutally detained by federal law enforcement agents for writing in chalk on a city sidewalk (but that's a story for another time)—I started hearing reports of people getting violently ill downtown from the chemical weapons the police were using. Causing more than the "normal" (albeit intense) response to tear gas and munitions smoke, the substances were debilitating and their impacts extended well outside the protest grounds. Something novel was being used, and people were rightfully scared. Were the feds adding new chemicals to their weapons of brutality?

An aquatic biologist by training with expertise in environmental toxicology, I knew where to start looking, and soon I was staring at the safety data sheets (SDSs) for weaponized chemicals like 2-chlorobenzylidene malononitrile (CS gas), capsaicin (OC spray), terephthalic acid (TPA), and hexachloroethane (HC). A regular part of "tear gasses," CS and OC induce acute but supposedly temporary effects to corral protesters, whereas TPA and HC are munitions smokes designed to signal or block lines of sight, yet they were all being dispensed similarly from handheld pyrotechnic grenades to blanket protest areas and parks, covering whole city blocks in a stinging, thick, unbreathable air.

As one might expect, the SDSs include very technical information about the human and environmental health risks of these compounds and explosives. But, hidden among the severe health impacts to humans, such as endocrine disruption and asthma induction, each of them came with the stark warning, "Toxic to aquatic life with long-lasting effects."

Because of their high heavy metal concentrations, these weapons are not to be used near groundwater or waterways. I was immediately worried because the main protest areas in Portland still use the ancient stormwater system, and these chemical weapons would then drain from the street directly into the Willamette River and into vital habitat for culturally significant and endangered salmon.

Although at the time I was barely able to get out of bed and take care of myself, I firmly decided that I, a supposedly "muddy boots" ecologist who has been behind a computer for most of the last decade, *needed* to start fieldwork. It was imperative to understand and halt the impact of these weapons on the aquatic ecosystem and, equally importantly, on the people standing up for racial justice who were the intended targets of these weapons.

In 2015, after previous experiences with inaccessible and unsafe work environments, I had started my own research company. So, as my own boss in July 2020, I was able to give myself the grace and time to recover from the assault. It was also clear, however, that once adequately healed I would need to use my professional independence to document these chemical weapons being used during protests—in real time—and study their long-term impacts on people and the environment; I needed to get up close and personal with these chemicals. In true "field ecologist" form, I took a rather "all-inclusive" approach to documenting this chemically aided brutality, taking countless photographs and videos of their deployment; collecting air, soil, plant, and other samples; and enumerating the large number of deployments.

There was just one major concern:

I rely on my service dog Wallace to stay alive and safe at all times.

PTSD, dissociative disorder, and severe anxiety make it extremely challenging for me to navigate social situations, and so Wallace accompanies me basically everywhere I go, providing stable support through specialized tasks like retrieving meds, engaging in deep pressure therapy, and providing a welcome outlet for stimming behavior (OMG his fur is so soft and curly!).

As one might imagine, however, chemical weapons are toxic to dogs. A small dog will die, a midsized dog will have life-altering seizures, and even a large dog will be debilitated from eating a capsaicin powder ball off the ground. OC spray is actually sold as a common rodenticide, but the police had spread it all over, leaving balls of it behind.

There's no protective gear that keeps a dog clean and safe. What does exist is lackluster and not something I would trust Wallace's (and thus my own) life with. The potential risk of these weapons to Wallace and then to me was a major factor that kept us off the Portland streets prior to 2020, where local police are notoriously fond of using tear gas and brutalizing disabled protestors. I only have to reflect on the days just after the election of Donald Trump, when Portland Police Bureau officers shot extensive tear gas and smoke into crowds of protesters downtown.

Having witnessed law enforcement in Portland in 2020 indiscriminately use chemical weapons without warning, target dogs with pepper sprays and projectiles, and fail to clean up toxic residues (including unspent powder balls) for months, I knew I needed a plan to keep both of us safe, and I needed help from some friends.

Thankfully, I have awesome friends.

Friday, January 22, 9:03 P.M.

[Me] Hey bb! What is your current involvement with direct actions that are happening?

[J] None atm, I'm pretty busy with houseless outreach. Why do you ask?

[Me] Okay, that's actually perfect. In some random hopefully never happens situation would it be ok for me to leave your phone number with Wallace when I go out?

[Me] Or like in advance of that I would do it all the time, you know what I mean.

[J] Yes absolutely.

[J] And _____'s if you want! We keep different hours and she's awake a bit later.

[Me] tytyty so much. Both of u.

[J] _____ says she's way more likely to be awake later, so def put both down.

[Me] I don't know if you've been following what's been going on, but I've been on Fox 12 and KOIN and the Feds have a nickname for me and it appears that they might be trying to go after me. I just want to make sure I have a backup in case s*** goes sideways.

[J] A nickname???

Over the ensuing months, as my car became a field research vehicle, the hatch turned into a mobile lab and the backseat became Wallace's second (or third? no, fourth) bed, complete with multiple water bowls, many snacks, comfy blankets, and snuggle buddies alongside a pack with his protective goggles, mask, and poncho.

It also had a letter sitting on the seat reading

This is Wallace Alfred Russel Simonis (he/him/his).

He is a six-year-old labradoodle, with up-to-date vaccinations (see attached documents).

He is the certified and trained service animal of Dr. Juniper L. Simonis (they/them/their).

It is against the federal Americans with Disabilities Act to separate Wallace from Dr. Simonis.

Wallace stays in the car while Dr. Simonis is working because it is not safe for him to walk or breathe in the areas where chemical weapons have been deployed.

Dr. Simonis regularly returns to the car; Wallace is well taken care of and has plenty of water and fresh air.

If there is a need, Dr. Simonis can be contacted using their cell phone at (___) ___-___.

In the case of emergency where Dr. Simonis is, e.g., unconscious, please contact

_____ _____ (she/her/hers) (___) ___-___

_____ _____ (she/her/hers) (___) ___-___

_____ _____ (she/her/hers) (___) ___-___

_____ _____ (she/her/hers) (___) ___-___

Thank you very much.

Although I developed a routine during the seemingly nightly barrage of chemical weapons to park a safe distance away, stay with Wallace until I needed to work, and regularly check on him when I had to be away, I did not take anything for granted.

Wallace and I were regularly separated at some of the most stress-inducing and dangerous times, when he was in the car while I was stuck in place documenting police brutality, diligently counting and photographing the chemical and physical assaults on protesters. If I moved from where I stood, I would be targeted and arrested simply for documenting.

Doing the science necessary to shine a public light on the impacts of chemical weapons put us both at substantially elevated risks. Working without Wallace by my side meant that I had a high risk of dissociating or having a severe anxiety attack that could render me incapable of keeping myself safe at a time when safety was paramount. And leaving Wallace alone made him vulnerable to being attacked and hurt by police.

Indeed, on more than one occasion, federal law enforcers (who knew me and Wallace from having detained us in July and from my public documentation of police brutality since) took advantage of us being apart to target both me *and Wallace*. They shot burning chemical weapon projectiles at or under my car and attempted to engulf it in toxic fog even though we were parked away from the protests, forcing me to stop working to move the car and keep Wallace safe. I could hear my dog-mom rage and fear recorded on the body cam video that kept rolling: "It's OK, Wallace, we just need to get to a new spot, it's OK, it's OK."

Clearly, it was not OK, and neither was I.

And as much as I just wanted to go home and snuggle with Wallace away from all the threats, I was almost equally determined to finish collecting spent weapons and air samples to send to the analytical chemistry lab for identification.

After every trip to document and study, when I returned to the car and decontaminated myself with a parking lot bath of water and wipes followed by a change of clothes, Wallace greeted me with a warm smile and good boy kisses, and I know if he could talk he'd tell me how proud he was of me; instead he snuggled that message into my lap at home as we slowly tried to unwind from the night's traumas.

Following extensive public pressure from a broad community coalition, including my first press conference and my first public testimony to a political body (Oregon State House), law enforcement ended the seemingly routine deployment of chemical weapons in Portland by late fall of 2020. But they kept sprays, gases, and smokes in their toolkits, ready to blanket downtown in tear gas and HC and to brutalize protesters demanding justice after police shot and killed disabled individuals in April and June of 2021.[1] And months and years after the summer of 2020 with its "Tear Gas Tuesday" and "Fed Wars," their impacts linger, visible on the stunted tree limbs and scarred grounds in public parks.[2]

I continue my work to document and understand these chemical weapons and their impacts on the environment, and I have shifted to studying the lingering effects in places like the Willamette River and public parks all over the city, where, thankfully, the air is generally clearer and there is no imminent threat of police assault.

But the spaces are still not safe: lying in wait in the soils, grasses, and waters of the city are untold unexploded and partially exploded bombs, each with the potential to seriously injure Wallace or other creatures or people. And so, for now, we stick to our nighttime ritual during the day and Wallace stays in the car. I hope that won't be the case forever. I want so much to walk in a protest with Wallace at my side, feeling that we, and those around us, are safe.

NOTES

1. Maxine Bernstein, "Portland Police Fatally Shoot Man in Lents Park," *Oregon Live*, April 2021, https://www.oregonlive.com/portland/2021/04/officers-responding-to-police-shooting-in-se-portland-park.html; Jennifer Dowling, "Tear Gas Used by Feds amid Protest Outside Hatfield Federal Courthouse," *KOIN*, March 2021, https://www.koin.com/news/protests/portland-protesters-protect-the-land-end-america/.

2. Mike Baker, "Federal Agents Envelop Portland Protest, and City's Mayor, in Tear Gas," *New York Times*, July 23, 2020, https://www.nytimes.com/2020/07/23/us/portland-protest-tear-gas-mayor.html; Alicia Victoria Lozano, "Federal Agents, Portland Protesters in Standoff as Chaos Envelops Parts of City," *NBC News*, July 2020, https://www.nbcnews.com/news/us-news/federal-agents-portland-protesters-standoff-chaos-envelopes-portions-city-n1234520.

Syreeta L. Nolan

18

THE DAY THAT CHANGED EVERYTHING

SYREETA L. NOLAN

Content warning: medical procedure, blood

This day began my first week of classes of the new semester, all on Zoom. I love and hate being on Zoom; it is not the undergraduate experience that I expected. I love not having to commute and being able to stay in my pajamas longer. First up is a meeting with Dr. Becky Petitt, vice chancellor of diversity, equity, and inclusion at University of California San Diego (UCSD), at 9:15 A.M. for thirty minutes. I love my meetings with UCSD leaders. They respect me as a disabled student and as an advocate for the disability community. We check in with one another openly as Black women in leadership and discuss logistics for our Disability Awareness Month celebration.

There's a fifteen-minute break before my first class of the day.

I begin Health Psychology at 10 A.M., the second item on my schedule. It seems fitting to take this class as a disability advocate. It is a normal first class: we learn about our professor, review the syllabus, and start to cover some course material. It is great to have a Black professor bring his experiences into the classroom. Fifty minutes and it's done.

I take a seventy-minute-long break before my next meeting to lie down with Hulu, watch a medical drama, and eat a snack. I have gastroparesis, so my stomach takes longer to digest food, and with stress comes nausea. I am supposed to eat four to five small meals per day, so I snack with the doctor's permission.

At twelve P.M. I log on to another Zoom meeting with Alisha Saxena, external vice president of UCSD's Associated Student Government; this is the third item on my schedule. Alisha is a nondisabled ally. I feel blessed to work with her because I am just getting used to my underrepresented student officer role with the University of California Student Association, and we are cofounding a Disability Ad Hoc Committee within this organization. I love the groundbreaking work that we do together. I was surprised when I was appointed to this role, as it is my first time in student government. Our thirty-minute meeting gives me energy for the rest of the day.

At any time, a flare of intense pain coupled with fatigue, nausea, and eating issues can take all my energy away and force me to cancel what I'm doing. I try not to get to that point by taking breaks and being gentle with myself.

I finally can lie down for a thirty-minute break. I eat something before bolting to my second class.

I log in to my last Zoom meeting of the day with Psychology 194A, Honors Thesis I, at one P.M., the fourth item on my schedule. This class is fifty minutes long, the beginning of my Psychology honors thesis year. I have successfully requested the Psychology Department to let me use my systematic review as my thesis. A systematic review is guided by a protocol and so is a reproducible analysis to understand the state of the field. The syllabus shows that this class is completely different from any of my other undergraduate courses. There are fewer descriptions of the breakdown of the grade because it is driven by my mentor and the norms within my discipline, Prevention Science. During class, I begin thinking, "I need to physically run to my appointments."

Next up is a required COVID test at two P.M. only ten minutes after my class ends, the fifth item on my schedule. I swear that one day I will see my brain matter or soul on this swab. Isn't this supposed to be just swabbing my nose? But the discomfort is worth it for the pain management procedures that I need.

I drive to get some lunch and then to my nerve conductance (a.k.a. EMG) appointment, the sixth item on my schedule. This day had space and grace leading up to my EMG, which is intended to verify my myasthenia gravis (MG) diagnosis that I have had since I was eighteen. UCSD Neurology is challenging my diagnosis because I have a mild presentation of MG.

Myasthenia gravis is a condition in which the muscles are weak due to problems with nerves receiving signals. My symptoms range from choking on my spit randomly to my ankles suddenly becoming weak to the point where walking becomes difficult. These are not the symptoms that doctors expect because a diagnosis of severe MG usually comes from an exam that confirms eye drooping, which I don't have. When this sign is not present in the office, an EMG and/or blood tests are performed to diagnose MG.

When I was eighteen, my first neurologist performed an EMG with a probe that discharged electricity into my leg. At the same time, the doctor pushed needles into my leg to force muscle contractions. The probe then measured how long the electrical signal took to travel through the nerve after those muscle contractions, evaluating their strength in response to a stimulus. As you might imagine, this test is painful under the best of conditions.

That EMG was negative and so was the blood test, but I was still having problems. The doctor was open to doing a trial of pyridostigmine, the medication used to treat MG, to see if I would improve. I am grateful that he was willing to do the trial because this medication improved my symptoms. Despite this medication helping me, I haven't been treated for my MG for several years because of the challenges by UCSD Neurology but today's short test could get me approved for pyridostigmine again. This will be my first EMG since being diagnosed with fibromyalgia and sciatica. I just want to get this over with.

I assume that this is just going to be another test as I check in at 3:00 P.M. for my 3:30 P.M. appointment. The doctors are a little behind, but that is fine because this has been a long day.

I sit there for thirty minutes wondering why I schedule things so closely, but what choice do I have? Because I'm a disabled student, my classes and my doctor's appointments compete for my limited time as I need breaks to recover from flares and my chaotic schedule.

After an hour of waiting, I go up to the desk and am told that I will be called back soon. Well, this gives me time to do some of my other work as cofounder of Disabled in Higher Education.[1] I review the interactions of our two releases that day— our "Disability 101" thread and the premiere of our video *How to Be a Good Ally*. This is the first week of our inaugural Disabled Empowerment in Higher Education monthly Twitter event. I am amazed to see a community coming together and the open conversations that are starting. As a team, we worked hard to bring this month together, and I am honored to be a part of it.

I am scrolling as I wait, loving, retweeting, commenting, and toggling profiles, making time seem to fly.

Finally, I am called back. A medical assistant gets my vitals and checks me into a room to wait for the doctor. I wish she could have given me a warning about the trauma and dismissal of pain ahead.

The resident doctor, who is new to me, asks if I have ever had an EMG before. I say yes, recalling my last experience in 2004. The resident breezes through the consent form; my intuition senses that he has skipped over some things, but because I am familiar with the procedure, I sign and then regret it instantly.

I knew this test was going to be painful, so I brought a small stuffed animal to get through the pain. While I lie on the exam table staring up at the ceiling, there are points during that test that I crush the poor thing as my mask muffles my screams. Left arm, right arm . . . strong electrical pulses. Right leg. Left leg.

Then he asks me to turn over onto my stomach. I am confused because every ounce of energy is being drained by each electrical pulse and a heavy brain fog is settling in. I hurt too much to question his command. The probe is driven deep into the back of my knee. I am on the verge of tears, and perhaps a few are escaping into the mask. This pain is now a 10/10 on their stupid pain scale.

I call myself out of the fog enough to ask, "Could you use less pressure?"

The answer is no, because then he would have to use stronger pulses.

I speak again: "I don't know if I will be able to walk out of this appointment. This is so painful."

He attempts to pacify me again, ignoring my pain. "You will be fine."

I am not fine.

"I am bringing in someone else for the second part," he says. I am angry and foggy at the same time thinking: "Excuse me!! What second part?!"

I am holding my breath between muffled screams and tears while at a stable 10/10 on the pain scale. I have no clue what is going on except that the nightmare is continuing.

"Turn on your side," he demands.

I am confused, with no energy to ask why. My right shoulder, where I experience most of my fibromyalgia pain, lies exposed. Electrodes are being taped to my neck. All I understand about an EMG is the electric probe and the needle tests; what are these electrodes going to do to me? An EMG is only a quick painful test, but this is turning into a marathon of pain. This person joining him is the supervising doctor, also new to me.

I think, "Where was she when he was torturing me alone?"

"Is this the part that he did well enough that he could torture patients by himself?"

"Has he had this procedure himself?"

He certainly should have had it, like cops who are tased so they understand the pain. I think, "Let's just find a cop to tase me and get your nerve conduction information from that. It is probably less painful and faster."

After they set up and he receives instruction, it is time for a series of five painful pulses in a row.

"Don't fight the movement." They don't need to tell me this; I have no fight left in me. I continue to give up as the time wears on.

First, a series of five painful shocks. The resident utters a halfhearted "sorry" when he accidentally discharges the series without warning. They really believe my pain tolerance is higher than it is.

I can't take this anymore. I work up the energy to ask about pain medication. The supervising doctor coldly tells me, "We don't give pain medication; it is not something we do." Why couldn't they have told me to bring my own pain medication for after the procedure?

I give up. I shouldn't have even tried to ask. I have nothing left. The supervising doctor occasionally pats me on the top of my head. I feel like nothing, am reduced to a dog trapped in a shock collar that keeps going off. I wonder if this is how people in the electric chair feel just before they die. I wonder if there is an end. The series of painful shocks moves to my arm. Pure agony. The first needle goes in with some warning, but all the rest go in without warning, like a cruel grand finale. No apologies. Maybe I am dead and that is why they aren't talking to me.

Maybe they don't think I can feel the pain. At my hip the needle is too short, so they switch to a longer needle. They are requiring me to do some movement with my legs with each new needle spot to activate certain muscles.

I am told, "You aren't doing it right; push harder."

"Turn your leg inward more," and I try my best.

They pull up the side of my underwear, say, "We're done," and quickly leave the room.

I lie there realizing that I had arrived there at three P.M.—it is now almost six P.M.

I think, "Wait . . . what is that wet feeling?"

I realize I am bleeding from my hip. Blood stains the side of my underwear now as I fight to stand. It is still bleeding; I need a bandage.

I pause to think, "Are they coming back with a wheelchair or to check on me?"

No one is coming. I use an alcohol wipe from the top of my torture instrument and find some bandages in a drawer. I must pause again. I don't feel good, but at least I am dressed now.

I open the door to the hallway to complete silence.

I look into the open door of the room that doctors use to take notes—empty. I struggle to walk toward the check-in desk. Maybe someone is there to help me to my car. I look toward the desk—empty. I pause at the end of the hallway in shock—no one is here. The clinic entry doors that were wide open are now closed. I realize that I am abandoned. I double-check that I have everything, and I leave. It is the hardest walk to my car I've ever taken.

As the abandonment washes over me, I think of when they asked about my scars while they were torturing me.

My scars are a history of years of self-harm and of all the mosquito bites I scratched. I realize that my humanity was not respected. Neurology had no right to ask about these scars while creating new emotional scars.

I have learned from my past that I must reach out to my counselor. I call and cry through a voicemail to my psychologist. I am able to schedule an emergency appointment for later that evening: appointment number seven. I keep crying. I realize that there is healing in tears, eventually . . . maybe . . .

I chose UCSD because my interventional pain management team is close to classes and my dorm. I never used to be afraid of new doctors, but I am now—I don't know what care I'll receive with the next doctor. All graduate schools I am considering will require changing my doctors. This EMG was a nearly two-hour nightmare of electric pulses and needles. It ruined my first week of classes. I frantically email my professors that night about needing to miss classes and reschedule assignments using my accommodations. I get together with Mom to eat dinner and cry in her arms. I go back to my dorm to sleep, crushed by the weight of abandonment, severe triggers, and the knowledge that my career and my life after graduation have been irrevocably transformed.

NOTE

1. @DisInHigherEd, https://twitter.com/DisInHigherEd.

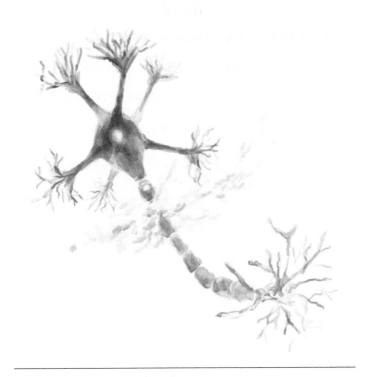

A misfiring neuron

19

BEING THE FIRST AND ONLY TO ASK

AMANDA O'BRIEN

Content warning: institutional ableism

"You're going to have to tell them."

A fresh flow of tears came as I realized that my dad was right. I was grateful to have my boyfriend to hold my hand because my parents were on the other side of FaceTime. "But it's your right. They legally have to accommodate this."

At this point in my graduate career, I dare not tell the department about my disabilities. I have lingering cognitive issues from suffering four brain injuries before my eighteenth birthday. I struggle with short-term memory loss, light and noise sensitivity, slurred speech, and aphasia, which involves difficulty with word retrieval. Sometimes the word I say in my head is not the word (or gibberish) that comes out of my mouth. Migraines exacerbate these issues by adding blurred vision, dizziness, nausea, and a halo clamp of pain and tingling.

By the start of college, I had learned that all my symptoms were in my head, and even if they weren't, I shouldn't tell anyone about my brain injury history. Final exams were always tough; if I had two exams in one day, I bombed the second, never scoring higher than a seventy-five, but I chalked that up to college. The first time I realized that there might be lasting effects of my brain injuries was during a presentation for a biology lab.

I was talking about giving aphids alcohol, or so I thought, when suddenly everyone looked concerned and confused. I could not understand why. My computer was working, I was running on time, and nothing was on fire. Then I started listening to the sounds coming out of my mouth. There were no words coming out, just garble. I paused, took a deep breath, and said, "Let me restate that." I picked up and finished the presentation, but I was humiliated. I didn't know how or why my brain or mouth had done that. No one said anything.

Nine months later, I met with a neurologist and heard, for the first time, that I wasn't imagining the effects of my brain injury and that I could take prescription medication to alleviate them. I was also told that I was eligible for accommodations. When I started graduate school, however, I had to decide whether I wanted accommodations. I looked around and immediately thought, "Nope." I was one in a cohort of nine, which meant that if I had accommodations, my professors and everyone in the program would know, and I did not trust them to react well. I already had the disadvantage of being a woman; I was not going to add my cognitive issues. There was no flexibility to the course load: we all had to take four graduate-level classes in each of our first two semesters. While my peers took on the challenge and made a run for straight A's, I knew that I could not work like that without getting massive migraines. My goal was simply to meet the minimum requirements to be in good academic standing.

Halfway through my second year, I began studying for "quals," or qualifying exams, which would allow me to become a full PhD candidate. In my department, we have to take four exams within two weeks. Each exam is an hour-long oral test in which the student speaks and writes answers on a chalkboard while the examiner, the student's advisor, and the committee watch. Each topic is based on a two-semester course sequence; any material from the course is fair game. Students take the exams in two sessions—two exams per session, back to back. The setup alone causes some healthy students to have panic attacks and nightmares for many weeks. My panic attacks set in several months before my exam dates, as I realized that this format was one of the worst ways I could be tested. While most students started studying about eight weeks before their exams, I started more than four months ahead of the exams, planning to take mine at the end of May. I had to build in time for migraines and bad days. I had to make sure I knew the material so well that there was no possibility of memory lapse on exam day. It was the only way I would pass. But as the migraines happened weekly, I realized that even these extra efforts might not enable me to perform at the same level as my colleagues. I would need additional help or accommodations.

Several panic attacks later, I came up with the accommodation: I needed to take the exams one at a time, split across four sessions instead of two. It was the same material, same time per topic, and all still taken within two weeks. My advisor, Professor Cory Lawson, was on board with whatever was best for me. Cory, however, is not a member of the department, so I still had to convince my department's director of graduate studies (DGS). Then I could go through the Disability Services office to get the formal paperwork, and then go through the regular procedure to set up my exams.

The first week of February, I met with the DGS to talk about quals and revealed that I was a traumatic brain injury survivor and would need specific accommodations. I asked, "If I get support from Disability Services, can I do this?" He responded, "Well, we're going to have to make an announcement and tell everyone exactly why you're doing this—because you have brain injuries." My dropped jaw gave away my shock, so he followed up with "I just don't want other students to think this is something they can take advantage of—like you have a good reason, but I want to be clear that there needs to be a good reason for this exception." I replied that I'd rather not reveal my brain injury history but that I could get the paperwork. He told me that we needed it before going further.

I finished the paperwork in two weeks. I also met with a different professor, Stefan, and asked him to help me keep my medical issues private. When I expressed concern about aphasia during the exam, he said not to worry—just be transparent when it happened and the examiners would give me the benefit of the doubt. Most days, when I wasn't in class, eating, teaching, or having a migraine, I studied from nine A.M. to eleven P.M. I began having trouble sleeping about two weeks before the exam and had several nightmares about failing, showing up naked, or getting kicked out. All the other graduate students told me that was common.

The first exam was the one outside the department, with Jim Baddick. I knew he would be the least merciful examiner. In class, he often made jokes like "This is obvious—sorry, I can't say obvious, that makes students feel bad—this is trivial," and everyone in the boy's club would laugh while I furiously took notes. However, he was the only professor who could give the exam, so I just had to endure this trial and the rest would be manageable.

The morning of my first exam, I woke up with my heart already pounding. By the time I was dressed and making coffee, I was sweating heavily. When I picked up the chalk, my Fitbit registered my heart at 110 beats per minute. Sweat streamed down my wrist. I fumbled the first question badly. It was a completely fair question from the textbooks about a concept Jim had referenced in the first week of class. The next question I struggled with at first, but I eventually got it. I know I did because the answer was on the board, but a different one came out of my mouth. He corrected the one that came out of my mouth. I told him that that's what I had said, and he said no. Cory nodded in agreement. I looked at the board and said, "That's the answer I meant to say. Sometimes I have aphasia." "Okay," Jim said, but I'm not sure he believed me. I nailed the next question perfectly. The rest were a mix: at least three times a different answer came out of my mouth than what was in my head. I tried to own when I genuinely messed up and when I had aphasia. Jim brushed off my statements with an "okay" or "yeah, sure." The hour finally ended. When I left the room, I knew my exam wasn't great, but surely it was enough.

When Jim and my advisor came out and said, "Let's speak outside," I knew what was coming. "You didn't pass. But you didn't fail." They said they didn't think I should do another exam but that they would "figure out an alternative." I had no idea what that meant. "I'm so, so sorry to do this to you," Jim said, "especially with all you've been through." Now I was fighting not only tears but also the urge to punch him for giving me a fake smile and weaponizing my past against me. "I hope this doesn't affect your other exams," he said.

The next day I received an email from Jim. "I know that nerves were an issue yesterday, but I am also sure that you really need to achieve a fuller knowledge of the material." How convenient to ignore the whole aphasia thing. I felt my insides turn to lead as I realized that the nightmare of quals could not possibly end this week. However, I couldn't think about it. I had three more exams.

I was numb for a week. I ignored texts or sent one-word replies while diving into Netflix binges and eating ice cream. I only left my house to mechanically do the motions for my other exams, in which I answered all but a few questions without help. I passed all three exams, but I still didn't want to have contact with anybody. My mom called and I lashed out and hung up. I knew I was being mean, but I did not care. I hated feeling so sick. I couldn't stop eating, even though I was never hungry. I couldn't sleep. I was ready to drop out of graduate school. I wanted to get a tattoo. Instead, I escaped for a planned vacation.

Two weeks later, I was told that I couldn't do research until I passed the exam because I was at risk of being kicked out of the program. My throat clenched, and I explained that I was confused: I hadn't failed, so I didn't need another exam, and I should be allowed to do research. The professor explained that for any other student it would have been a pass, but Jim said that because this was my research, I needed to learn the material and take another exam before I could continue. To adequately prepare, I would have to study eight to ten hours a day for two months, a feat I could not handle but had to do to stay in the program. Would they really kick me out? I had an external fellowship for three more years and had passed the other three exams.

I went to Stefan's office and laid out the entire situation. After ten minutes of silence, he finally told me that this whole thing was a department handbook violation. There were two options in the oral exams: pass or fail. Pass, and you're done. Fail, and you must retake the exam within a month. The topic can be narrowed in scope, but only material from the first exam is fair game. If the examiner is borderline on whether to pass or fail, the decision goes to the advisor. He took a breath and said, "Go home." He'd run interference with everyone. I felt relieved that I was finally being heard but was more confused than ever. All I could do was go home to my apartment and wait for news.

Four weeks after my terrible exam, my department determined that I would only need to do a problem set from Jim to pass. For the next two weeks, I worked on the problem set and did research. When I finished my first pass at the problems, I video-called Jim for questions. He ended with "Again, I'm so sorry to do this to you." I thanked him and signed off. Since I couldn't punch him, I did the next best thing: I proofread my completed problems and submitted them three weeks early. That night, Jim emailed me: "Okay, you've passed your prelim now. . . . It was mostly correct." There was a sentence about what the errors were, a congratulations and best-of-luck sign-off, and that was it. I felt nothing, just numb. I didn't want to celebrate. It did not feel like a victory or a relief. While they walked away and moved on with their lives, I was left to deal with my injuries.

I did not physically recover from these exams until I went on medical leave more than a year later. During that time, I learned of two conditions related to my head injuries that had gone undiagnosed for nine years: neck arthritis, which exacerbates migraines, and convergence insufficiency, which means that my eyes did not track together. Considering everything, I realize now that my qualifying exams were not just a mental and emotional stressor but also a physical one. If you tell a professor to measure a graduate student's ability by the time it takes to run a mile or how much weight the student can lift, the professor will scoff and say these actions have nothing to do with research. Somehow, though, the same professor will justify the qualifying exam system as a measure of research potential, when in reality it's glorified hazing, ripe for discrimination and abuse. Meanwhile, there has been little to no research on education outcomes with these exams.

There's no reason to keep the same archaic exam system in place when no two programs have the same practice and some programs have replaced this system with a research proposal. There's no reason not to offer flexible options to spare students the need to secure formal diagnoses and pages of paperwork from their doctors on top of the rest of their graduate studies. As a community, we must realize that these exams are completely unnecessary and bring our pedagogy back to preparing students for research.

Stephanie Schroeder

20

LOST IN NEW ORLEANS

STEPHANIE SCHROEDER

Content warning: n/a

As I walk slowly, my ankle begins to stop functioning. I am all too familiar with this symptom: I am no longer able to flex my ankle after walking a mile because my multiple sclerosis (MS) gives me temporary foot drop. I did not wear my foot brace today, and my ankle is responding. I have to focus on picking up my leg using my knee or I will trip because I can't lift the front of my foot. At the same time, I notice that the lanyards around people's necks are purple, whereas they are supposed to be blue for the 2016 Ocean Sciences Meeting (OSM). I am lost in the huge New Orleans convention center, a building that spans nearly eleven blocks, and I wandered into the wrong conference. Because I've told few people about my diagnosis, I'm not sure if I can reach out to anyone and explain that I am simply having difficulty walking. I've immersed myself in the conference by attending talks and poster sessions, but what I really want to do is sit down and cry. This is my first time navigating a conference with MS, and I am exhausted, not just physically but also mentally and emotionally. I am struggling with my disease because I have not reached out to people, even though I witnessed the pain and trauma my mom went through as she dealt with her own MS.

My childhood was filled with camping, hiking, and the whole gamut of outdoor activities. My parents were both teachers, and our summers were spent traversing the country in our wood-paneled station wagon, exploring as many national parks as we could. My mom started experiencing strange symptoms like tingling arms and legs in the late 1980s. I was eleven years old and oblivious to her internal struggles: she was not diagnosed with MS for years and was told that her symptoms were all in her head, but she noticed that she had a difficult time keeping up on bike rides. Worried about her employment as a kindergarten teacher, she endured her disease without reaching out for help from her coworkers or employer. I had a 5 to 10 percent chance of inheriting MS, but I didn't want to think about that possibility. I had watched my mentally strong mother face the physical limitations of her MS, but I was not prepared to deal with my own battle. Now confronted with my own struggle, I had opted to suffer in silence like my mother and initially confided in only a small group of friends and family. I was lucky enough to not show any manifestations of the disease and remained quiet about my invisible disability, not wanting to appear weak or to accept my own hidden bias against people with disabilities.

MS is an autoimmune disease that attacks the myelin sheath surrounding the nerves, which conduct electrical impulses throughout the nervous system, allowing the brain to efficiently communicate with the body. A myelin sheath acts as insulation around the nerves. When that insulation is damaged, electrical impulses don't always get where they need to go. MS affects people differently; some people are subjected to "cog fog," or cognitive impairment, while others are physically exhausted by simple tasks. I am in the latter category. On a normal day I can run (or rather, shuffle-jog) a couple miles in the morning, but on other days I am drained simply from waking up. I had considered the physical fatigue when planning this trip, but I hadn't anticipated the mental frustration from dealing with a chronic illness while attending a conference.

Just two years before, at the 2014 OSM in Honolulu, I had happily trekked a mile to and from the convention center from my hotel, enjoying the beautiful tropical island. This was prior to my symptoms and diagnosis. I had been stuck at the airport for ten hours waiting for my flight to Hawaii and finally arrived late at night to the hotel where I was staying. I managed to wake up early the next day, participate in an all-day workshop, run back to the hotel for a quick lounge by the pool, and then head out for dinner with friends. That was day one of the meeting, and I maintained this level of activity for the rest of the remaining five-day conference. I ran around, reconnecting with colleagues, going to mixers at night, and attending early breakfast meetings, not realizing that I would later find myself longing for these days.

I was diagnosed with MS in January 2015 and thought that within a year, or around the 2016 OSM, I would have adapted to my chronic illness. After emerging from the foggy haze of my diagnosis, I attacked the disease head-on, changing my diet, seeing a therapist, practicing meditation, and renewing my commitment to regular exercise. As a proactive person, I believed that I had fully accepted my new normal, which included assessing my energy budget and mental capacity each day. Healthy people normally do not have to think about how every action will affect their energy level, but now I constantly think about how a day's activities will affect my mental and physical state. If I run two miles in the morning, I have to recharge for at least a couple hours before I can walk two miles in the afternoon. Recharging means I need to sit and let my body recuperate while my brain continues working, as even a short walk around the block is difficult.

After I was diagnosed, I retired from a full contact sport that I loved, roller derby, because it requires a tremendous amount of balance. I had taken up roller derby in graduate school to forget about the stressors of academia. Putting on skates, sailing around the rink, and participating in a full contact sport helped me escape from experiments and literature searches. Admittedly, there was also some bloodlust involved, and I loved being able to take out my aggression on the rink. I took for granted how my body grew stronger and I was able to improve the more I skated.

Now I recognized that, despite lots of physical therapy, being on wheels was too much coordinated activity for my chronic challenges. MS messes with my balance, and sometimes I have to focus on simply not falling over; balancing on one foot has become a challenge, when only seven years ago it was easy. Skating around a rink while avoiding or hitting other people was now too dangerous; I was a liability and was asking for a broken bone.

I thought I was coming to terms with my new normal, taking up the accordion and giving up skating, but it's one thing to quit an activity and another to have to assess one's energy budget every day. I was not mentally or emotionally ready to accept that walking more than a couple miles without my ankle giving out was the new normal, no matter how much I pushed myself.

Conferences are overwhelming and exhausting, even without a chronic illness. Research talks are back to back, and attendees must run from room to room every fifteen minutes or less, all day long, and then look at scientific posters for hours. Then they spend the evening talking at happy hours with colleagues from across the country. The events that happen outside the conference are often as important as the conference itself, and skipping them could mean a missed opportunity. Networking at conferences can lead to collaborations or even job opportunities, and there is not a moment's rest. Attendees spend a lot of money for conference registration, travel, and hotels, which hopefully their institution has paid for, and they want to maximize their time.

At OSM 2016, the New Orleans convention center was massive. I had to walk down a very long hallway to get from one section of the building to another; it felt like the length of fifteen football fields but in reality was maybe a quarter mile. To attend all the talks that interested me, ranging from rocky intertidal ecology to marine education best practices, I needed to run up and down that hallway multiple times a day. Even though I had conserved my energy by not exercising in the morning, having to essentially sprint from eight A.M. to six P.M. gave me no time to recuperate and rest.

I hadn't considered all the factors that would drain my energy during this trip. I was on medication that made me sleep poorly, I was walking more than my body was prepared for, and I was still trying to adjust to a new diet: my neurologist had suggested that I give up dairy and fatty foods to help support my immune system. Beignets, etouffee, gumbo; decadent Cajun food native to New Orleans: all of them were now off my list. As I searched for breakfast one morning at a nearby coffee shop, I was told that it didn't have anything dairy-free but that the muffins were free of gluten. I grudgingly bought coffee and a banana, annoyed that the barista was ignorant about dietary restrictions, and hoped I could find a place that carried oatmeal. There were too many variables to control. I was staying at a hotel that was half a mile from the convention center because I thought the walk would be enjoyable. I didn't realize that adding a mile every day to my trek would wipe me out physically. I had ordered a foot brace from Amazon to counter the foot drop; the foot brace helped, but it did not relieve my exhaustion. I wasn't ready to share my illness with friends and colleagues at the conference, so I continued to wear an able-presenting mask. So that no one would question my abilities, I wanted to appear "fine," as if I was tired from the conference, not completely drained every day due to a chronic illness.

Eventually I was able to calm down after my initial panic of getting lost in the New Orleans convention center. Swallowing my pride, I sat down on the floor to conserve my energy, consulted a map, figured out where I was, and navigated back to OSM. But it was not a great way to start a conference and set the tone for the rest of my trip.

As I approached OSM 2018, held in Portland, Oregon, I used more strategic planning. I studied the layout of the Portland convention center to ensure that I wouldn't get stuck in a random, faraway section and scoped out quiet areas where I could sit down and take breaks. I did not go on walkabouts between sessions unless I knew exactly where I was going. Three years after my MS diagnosis, I was beginning to comprehend what I needed to do to balance my energy budget and manage my emotions. I selected a hotel across the street from the convention center that had a café so I could easily grab breakfast. I also checked that it would cater to my dietary restrictions. I parked myself in conference sessions and did not go racing from one talk to another every fifteen minutes. I said no to events. I went to bed early and woke up early to exercise but made sure to factor in walking to and from events. I invested in an ankle foot orthosis, a custom-made carbon fiber brace that aided in walking. I was mentally and physically in a much better place, listening to my body and truly recharging at night. I was selective with my networking activities, opting for more meaningful connections, meeting with only a few people, and setting up informal get-togethers at nearby coffee shops before the conference that were away from the frantic pace of the convention center. And I opened up my support circle and confided in more friends; some people at the conference knew about my MS, and I could turn to them if I needed help.

Getting lost at OSM 2016, both literally and figuratively, was the reality check I needed to acknowledge my disease. As much as MS is an invisible disease for me, in stressful situations it becomes apparent that it is not invisible and my body and brain tell me so. I still get frustrated that my body cannot perform what were once simple tasks, like running two miles, but I've worked hard to change my mindset and focus on what I'm still able to do. Although I still struggle, I've come a long way from my experience in the New Orleans conference center; I'm prepared to reach out for help or sit down and recharge. My new normal is a moving target; I can never predict what's going to happen each day. But at least now I have the coping mechanisms and mental capacity to understand which direction the moving target is headed.

Divya M. Persaud

21

ASCENDING THE CINDER CONE

DIVYA M. PERSAUD

Content warning: n/a

In the middle of a drought-baked summer, a rain cloud lazed over Mount Lassen; a few drops came as I wandered down the road to a patch of concrete and eased myself down. A lava flow was to my left, its edges blurring under the smudge that was beginning to thunder in the distance. To the right were brush and boulders and beyond that the forest. There was nothing else. Maybe there was a flap of wings farther down the road, then nothing again. Lassen is a place of solitude, especially here by the radio observatory. I could hear a breeze ten miles away, whispering through the trees seconds before it arrived at the branches above; vultures moved as if they were leaves themselves. Roads were empty and hot. And at night the light of the stars revealed the shapes of the trees.

It was on this patch of concrete where days before I had first fully taken in the Milky Way. After years of hiding in the shadows of cities, here it was in its totality. My astronomer friends pointed out individual stars while I felt my body sink onto that concrete. As we wandered back to the lodging, our minds back on Earth, we had to hold one another as gravity embraced us again. Our laughs traveled down the mouth of the road, its jaws burned evergreens interspersed with survivors, and disappeared.

The emptiness was enormous in my stomach and was only growing. Unlike my surroundings, I wasn't sure what I wanted; I could only look at this cloud as it crawled forward. It was easy for the cloud to just *be*. It was easy, too, for the lava flow to just lie in front of me, to lay bare the tephra, the cinders left over from frothy lava, the burnt brush from recent years echoed this ancient time. It was all in front of me: a map of an unknown place, of unknown times, the answer to a question I didn't know.

My route had once been simple and ravenous. I had chosen to be a geologist at four and a planetary geologist at eleven. The path to these roles was clear by thirteen: secure a NASA internship, learn what research is, enter a degree program in geology, and, if all of these steps worked, pursue a PhD in planetary science. So my foray into research at sixteen with an instrument team on an active mission did not whet my appetite as much as give me a deep ache of hunger in my stomach. And I wanted to be a published writer and a cellist as well. There was only hunger.

When, after a brief viral infection in October of my sophomore year of college, I realized that the soreness and fatigue weren't going away, this vision started to crumble. Walking to class was difficult, and so was holding a pen, standing, eating, or using my phone from bed. Something unknown raged through my muscles and left them tired and unwilling, while each strand of every muscle felt as if it were being pulled from above and below every other minute, as in a marionette. That academic year did not last long; in January the fatigue was too great, and I chose to take a medical leave for the semester with the hope that I would return later that year.

This internship in California was a glimmer of hope in the summer—the possibility that my body might be repaired enough to feel that hunger again. In the van to this field site, the sight of the alpine forests as we ascended thousands of feet created only more of this hunger. This experience, I hoped, would be another taste of what I needed to be.

But in Lassen my body was as sensitive and vulnerable as ever. There were days of hikes over the first volcanoes I had ever seen, through ash-covered forest floors and along hills colored like spilled oil in water. Two decades spent hungry to see this planet weighed on each of my steps as I refused to rest and look away, while in all of this quiet there was no phone signal to call my doctor about the new ways my body hurt. My stomach began searching for nourishment I couldn't give it, searching for something, perhaps, to destroy, and I was disappearing. But that real, other hunger was also visible and underfoot. I wanted to bury the word "illness" in the crunching of my steps in the volcanic soil.

One of the hikes had begun with the ascent of Cinder Cone, a volcano whose remnants were only broken rocks. Our feet sunk into the crackling, fist-sized cinders, and soon the height and heat clutched my chest and made me stop on the slope. In the distance, around this little reddish scar, were miles of deep forests and, beyond, more volcanoes capped with snow and purpled by the haze. Everything was close and yet so far, and the wonder threatened to swallow me, even as I struggled to breathe.

A friend ahead of me slowed down; we linked arms and slowly wound our way up the mound. At the top was a bowl filled with trees, and we rested on the lip to lunch. The fantastic scene dazzled me, and joy pressed against my chest. But my feet ached, and I was dizzy. This dream might kill me, if not this week, then eventually. I was doubling my lupus medication, holding my body in front of the mirror and feeling its growing absence, fearing that this disease was swallowing me bit by bit every day and then eventually.

On that patch of concrete I let it all sink into me. The humidity and elevation were thick in my lungs. I wasn't sure who I was if I was *unable*. If I had to turn my back on my dreams, would I still be the same person? Was there a lesson to be found in my illness, some moral in the aches that I needed to embrace to find the answer?

Mist from the rain cloud above the mountain moistened my face as I stared at it. I closed my eyes and thought of the week. One day three of us had taken a detour to avoid the hikes and found, tucked away in the contours of the volcano, a spit of turquoise shimmering in the heat. The lake was quiet, but in that incredible strip of color, the stones caressing it and oozing yellow sulfur, I felt I could breathe a little. The mist was soft on my skin.

When I opened my eyes, it was all still here: the quiet, the cloud disconnected from other clouds, hovering over rock and brush. Arms around myself, I thought of the hares and deer combing the baked pine needles and the vultures that waited for nothing at all. I thought that Lassen was a marriage of extremes: heat, pressure, explosion; the massiveness of the Milky Way; the unfolding of new life from burned trees; the microbiota in bubbling muds; the exposure and seclusion. I was held on the warm concrete, and I couldn't look away from the rain.

I had always wanted to be something. Since the onset of that flu that felt like it had never ceased, my resolve had whittled away each day, as especially my hands—my resolute companions in all of my endeavors as I urgently tried to get what I wanted—told me they could not be what I wanted. But the water emerging from the quiet mist and the basalt laid out beside me, her spine warped but at rest, her fingers splayed and broken—they weren't meant to *be* anything. Rock just is. The sky, the vultures, and the winds that announced their arrival from miles away just *are*.

The rain just happens. There is no ethos to be extracted from the way the volcanoes rise above this landscape, and none from my pain, either. To exist is to do just that.

Just be, they said.

I didn't know how I would continue to crumble in the future; I didn't know the shape I would take or what else might fall apart, but I knew, suddenly, that I was torn between a sense of dying and living, gripped by the question of what the future would be: that I could, at the very least, let myself exist.

I learned to be a scientist at ten thousand feet of elevation. Our identity emerges from our scars and the ways that we heal: the moments of care we offer our bodies and minds when we might otherwise take too much from them. A lava flow might suture a landscape; a forest might regrow from its own devastation.

I stood and returned to the lodging, where my friends were looking at fluorescent minerals in the dark. They beckoned me to join them. Our talk and laughter rang out into the woods.

V

REFLECTIONS IN
THE WATER

As water reflects the face, so one's life reflects the heart.
—Proverbs 27:19

We cannot see our reflection in running water.
It is only in still water that we can see.
—Taoist proverb

When events happen in our lives, their impact, shape, and meaning change as more happens, or our perspective simply changes as we continue to reflect on these events. The authors of this section have experienced events in the past—some acute and some protracted—that stretch over long periods of time with multiple diagnoses, surgeries, or medical investigations.

Glyn, a sociologist who uses a wheelchair, begins this section with a story about lessons from a fieldwork experience in the United States and imagines what might lie beyond the "social model of disability," especially in the field of public understanding. We then meet Emma, who comes into graduate school with a traumatic brain injury and reaches out for help in many ways but comes up against roadblocks and has to rethink her entire career. Leehi embarks on her first field research trip to Greenland, where she starts to experience new physical pain and the grief associated with climate change. Katie shares her history of alcohol and mental illness in graduate school, experiences that feel like an endless and expanding abyss, especially in the absence of support. Sunshine is a midcareer academic who is diagnosed multiple times with cancer and who must use her scientific mindset to understand what is medically happening to her. Olivia closes this section with vivid descriptions of nighttime fieldwork and moments at home as she moves beyond the tidy columns of data that once ruled her life.

These pieces look back with hindsight and muse about the broader implications of privilege, access, ableism, and adjustment to the scientific community with a disability. Many stories deal with grief, but there is also joy, acceptance, and, above all, hope.

Glyn Everett

22

THINKING BEYOND THE
"SOCIAL MODEL OF DISABILITY"

GLYN EVERETT

Content warning: n/a

I n 2012 I started a job at the University of the West of
England, Bristol (UK), looking at a strange and curious new
thing for me called "blue-green infrastructure." This name
refers to attempts to manage urban flood resilience by bringing
together and working with urban green spaces (parks, gardens,
rain gardens, green roofs, etc.), naturalised waterflows, and water
storage—ponds, wetlands, deculverted rivers, rain barrels, and
permeable paving. More green space in our cities has to be a
good thing—I confess, the hippie in my heart exulted at the
thought of researching this subject.

The project's principal investigator brought our team to
Portland, Oregon, in the first year. We spent a week seeing the
good blue-green infrastructure work that had taken place in
Portland since 1996. One year later, we returned for five weeks to
look in more detail.

On my first night back in Portland, after a thirteen-hour
flight from Bristol, England, I had to go straight out to see one
of my favourite bands, Moon Duo. I am a manual wheelchair
user and have been for fifteen years now, and at this point I was
very impressed by the power of the Americans with Disabilities Act (ADA); I hadn't had the chance to review the venue's

accessibility ahead of time, but it turned out to be roll-in, all level, and with an accessible toilet. Thumbs up for access so far.

I am here to see blue-green infrastructure, though, not concerts. I am in Portland to study public attitudes towards and awareness of bioswales, which are highly engineered rain gardens using native plants. Bioswales are designed to take excess stormwater off the streets and then remove heavy pollutants from the water before it returns to the rivers. In Portland, the Willamette and Columbia Rivers were heavily polluted from industry and road runoff, and the salmon and trout populations had dropped drastically. With the installation of these water quality and flood risk management measures, over the past fifteen years the water quality has improved and fish populations have been recovering.

In Bristol I had done this sort of research by knocking on doors, distributing leaflets, and conducting interviews in people's houses. With almost all the houses I approached, I could roll up to the front door and knock or press the buzzer; there might be one step up into the house, but I could usually flip this myself or get in with a helping push from behind. But in Portland most of the people I wanted to interview lived in houses with four or five steps up to their front doors. I could immediately see that being on wheels would pose a problem.

On our first trip to survey the areas, of the six streets we had identified for researching, four had inaccessible houses where chatting at the door was impossible for me. That evening, I retreated to our motel to sulk and to consider reframing or reorienting my work over the next month. The "social model of disability" was at the forefront of my mind: the argument that we, "disabled people," are not innately disabled but rather are disabled by society. We may have a variety of *impairments*, but it is the attitudes, expectations, infrastructure, and provisions of society that actively dis-able us, create problems, and affect our social functioning.[1] This was a powerful argument and served as a foundation for campaigning for the UK's first Disability Discrimination Act (1995). And it had never felt truer than now. I knew that I was perfectly capable of conducting interviews and gathering data from people, but the ableist architecture of Portland left me unable even to approach people in their houses and ask them to participate.

I wondered about asking one of my colleagues to come out with me to knock on doors, but we all had our own distinct research programs that were demanding of our time, and I would have felt terrible taking others away from their work. I considered placing signs in local papers and mom-and-pop stores asking people to call me. However, from my previous work in England, I knew that the response rate would be very low, and I had a target number of interviews to achieve. Although it wasn't my fault, I still felt embarrassed and downhearted about my first international research project coming to such an unsuccessful end so soon.

After a long night of feeling very down about things, the next day I spoke with my colleagues. Nobody had any idea how to resolve the problem (other than, as mentioned, taking time out of other people's work), until I mentioned it to the principal investigator. He told me that we were to be joined by a Hong Kong colleague currently based at the University of Nottingham, England, and that he had no specific role ascribed to him. I was overjoyed to find a colleague with free time and functioning legs who could knock on doors and get people out for me to meet. Whilst this didn't overcome the problems with access, it certainly promised to overcome issues with conducting research. He could be my "enabler" or personal assistant (PA), a common phrase in the UK.

The next morning I met my new colleague, Faith, a lovely and friendly chap. I explained what I needed him to do, somewhat fearing that as an established researcher he might take offence at being reduced to knocking doors, but he was enthusiastic about enabling me to do my work. We set off after breakfast and had a productive day that was blessed by an unusually sunny May (especially for the Pacific Northwest). Faith would knock on doors, introduce himself, and ask householders if they would mind walking down to the street to talk with me. I don't know if the sight of an Englishman sitting in a wheelchair at the bottom of their stairs tweaked a few heartstrings or if a keen and smiling Hong Kong chap brought some sunshine to their day, but we had good success in gathering willing respondents to talk. I felt once again back in my zone and went ahead gathering good data from people about their awareness, understanding, and preferences concerning the bioswales (which it turned out weren't that great, as I have explained elsewhere).[2]

We continued like this for about a week, trekking the streets with Faith knocking on doors and me conducting interviews. Then one day we visited a new street that was level-access to all the doorways. I was excited because I would be able to do my own introductions—a small matter, but one of personal pride. However, at the first door I knocked on, I was greeted by an Asian gentleman speaking a language I didn't know. I felt awkward and embarrassed but of course unable even to apologise in a language he could understand. I began to back away, bowing my head to indicate an apology for having disturbed him, but before I had moved more than six feet, Faith began talking to the man. The gentleman responded, and they were soon engaged in a full-hearted chat. I left them to continue for five minutes or so, and when Faith returned his attention to me, I stared at him in confused admiration. He then explained that this entire street was primarily occupied by recent first-generation immigrants from Cantonese-speaking parts of China who were working in and around the many Chinese and Asian restaurants and wholesalers of Portland. Faith, being from Hong Kong, spoke Cantonese as his first language.

This was a most unexpected turn of events; we had finally found the street where I could approach the houses directly, unimpeded by ableist architecture, and I was instead impeded by my inability to speak Cantonese, whereas Faith, hitherto reduced to knocking on doors, was now in his element. I asked if he would be willing to take on my role, and he gave me his biggest Faith grin—he'd love to. So I quickly ran through the interview structure and objectives with him and then took a backseat for the rest of the day, wheeling up and down and observing the bioswales and people. Faith did some marvellous work, encouraging engagement from a group of people who work very hard for long hours and who, as it turned out, had felt ignored and excluded from the city's agendas; they could with good reason have chosen to reject any opportunities to participate in our research.

These interviews revealed that most residents were completely unaware of the purpose of the bioswales because the literature sent around by the city was written solely in English and these hardworking people did not have the time to learn another new language. They observed the green areas by the sidewalks along their streets but did not know if they were allowed to even touch them, to remove weeds, or to water them during dry periods, and so the bioswales acted more like an imposition than an improvement. Many appreciated the additional green space, but they were disempowered and disengaged.

This made clear to me a much wider sense of the "social model of disability" principles: that people can be prevented from understanding and participating by *any* actions that do not consider the multiple different publics in a city. For the first time, I realised that the "social model of disability" could potentially speak to and for most of the world's citizens, not only the disabled. People can be socially "disabled" by much more than impairments. This is not to take *anything* away from the power and implications of the "social model of disability," but simply to argue that we also need to develop and frame a "social model of exclusion."

This awareness in turn brought home to me the ongoing relevance of Robert McRuer's *Crip Theory*,[3] which describes how many social institutions, infrastructure, discourse, and practices are "written" by straight, white, cisgender, heterosexual, able-bodied people and how the great majority who do not fit this minority "norm" can become excluded or disadvantaged. There is, it would seem, a strong ongoing need to *crip* the discourse and practices of much of the world—of city governance and of academia for starters, in my case—to unpack and flip the conversations in order to expose how narrow the intended audience is. We need to try to have multiple conversations, in multiple languages, so that we can produce multiple solutions to the "wicked problems" that we face.[4] To do otherwise is to encourage disengagement and discourage active participation, which could only increase our chances of failure in dealing with flooding, climate change, and all of our other global social problems.

Since coming back to England with my new and expanded understanding of the relevance and importance of the social model to much broader swathes of society, I have been involved with much more fieldwork, embedding myself in a variety of green spaces to understand more about public awareness, understanding, and preferences. I have discovered again and again how often and easily simple scientific communication can be

overlooked, leaving citizens as passive rather than active partici-
pants with the natural world placed around their developments.
At one site, developers installed ponds to reduce flood risk and
then moved on, leaving no information about the ponds' purpose. Local residents and the local authority community rep-
resentative had no idea of their function and so simply didn't
value them. At another site, developers planted an invasive type
of bulrush in a pond and moved on, leaving residents ignorant
of the fact that the bulrush was reducing storage capacity in
their pond and so was affecting its usefulness. Finally, at a third
site, permeable paving was installed on the streets of a neigh-
bourhood and no attempt was made to inform newer residents.
As a result, some residents filled parts of it with concrete, thinking they were doing the right thing to reduce the weeds whilst
actually reducing the streets' permeability. The outcome of all
these practices is that the flood management capacity of devices
is reduced, and so potentially are the proposed "multiple social
benefits" of using blue-green infrastructure, because residents
neither understand nor appreciate what is being provided or
how best to behave around these devices to help ensure their
longevity. I hope to work further on this issue, engaging with
authorities and practitioners to open up spaces for collaboration
and cooperation that can empower "local people" to have more
voice and agency, and hopefully to improve appreciation and
behaviour as a result.

I am even more keen to turn my practice towards "my own
people." How do different groups of disabled people interact
with blue-green spaces? What barriers do they face or feel? Can
vision-impaired people access and enjoy these spaces? What
could improve access? Could signage, design, or plant choices
improve the experiences and understanding of different neurodivergent people? These questions will need unpacking, and the

questions themselves will need reframing with different disabled people and groups. I have as much to learn as anyone and would never pretend to "have the answers" to such complex questions. This is where research needs to slow down, listen, and learn.

I also want to investigate and look more into disabled people's resilience. In the literature on emergency planning, we are often positioned as simply being "vulnerable" and so lacking relevant agency and expertise. However, I strongly believe that we disabled people, having had to deal with a range of limitations, negotiate all the barriers within an ableist world, and plan for our own upsets, limits, and emergencies, can be much more resilient than we are given credit for. Listening to disabled voices and respecting our embodied knowledge could help city authorities and urban and emergency planners to improve their work in planning for the future. And for myself, this means getting involved, getting research money in, and then teaching more and more future built environment professionals about how we "disabled people," at 20 percent of the population, are one of the biggest minorities in the world. Ignore us at your peril, for any one of you might join us tomorrow. Things are changing, so I won't say I'm not hopeful—we just have to keep pushing (and pushing) for more change.

Acknowledgments: The research was performed as part of an interdisciplinary project undertaken by the Blue-Green Cities (B-GC) Research Consortium (www.bluegreencities.ac.uk) with the Portland-Vancouver ULTRA-Ex (Urban Long-Term Research Area – Exploratory) project (PVU), as part of the "Clean Water for All" initiative (www.epsrc.ac.uk/funding/calls/cleanwaterforall). The B-GC Consortium is funded by the UK Engineering and Physical Sciences Research Council under grant EP/K013661/1, with additional contributions from the

Environment Agency and Rivers Agency (Northern Ireland). PVU was funded by the United States National Science Foundation award #0948983. We thank the city of Portland Bureau of Environmental Services, the Bureau of Planning and Sustainability, and the Johnson Creek Watershed Council for their generous support and sharing of data, field equipment, time, and expertise.

NOTES

1. Michael Oliver, *Social Work with Disabled People* (London: Macmillan, 1983).
2. Glyn Everett et al., "Delivering Green Streets: An Exploration of Changing Perceptions and Behaviours over Time Around Bioswales in Portland, Oregon," *Journal of Flood Risk Management* 11 (December 2015): 973–85, doi:10.1111/jfr3.12225; Glyn Everett et al., "Sustainable Drainage Systems: Helping People Live with Water," *Proceedings of the Institution of Civil Engineers—Water Management* 169, no. 2 (April 2016), doi:10.1680/wama.14.00076.
3. Robert McRuer, *Crip Theory: Cultural Signs of Queerness and Disability* (New York: New York University Press, 2006).
4. In 1973 Horst Rittel and Melvin Webber introduced the phrase "wicked problems" to characterise the complexities and challenges of addressing planning and social policy problems that lack clear definition or delineation of affecting factors and actors; these problems are irreversible and interact with and feed from and into other problems. Sustainability, for instance, poses a classic wicked problem. Vincent Blok, Bart Gremmen, and Renate Wesselink, "Dealing with the Wicked Problem of Sustainability in Advance," *Business and Professional Ethics Journal* (2016), doi:10.5840/bpej201621737; Horst W. J. Rittel and Melvin M. Webber, "Dilemmas in a General Theory of Planning," *Policy Sciences* 4, no. 2 (June 1973), doi:10.1007/bf01405730.

Emma Tung Corcoran

23

SUFFER IN SILENCE OR LEAVE

EMMA TUNG CORCORAN

Content warning: vomit

When I first began my doctoral program, I wanted to stay in academia and change things from the inside as a professor. But as I'm nearing the end of my PhD and finding myself constantly exhausted by the ableism I face daily, I realize that I won't be able to effect any meaningful change if I have to spend all of my effort fighting for equal access.

I remember the first time that I set foot in a molecular biology lab. In my freshman year of college, I took a lab-intensive biology course and was introduced to the minuscule world of DNA, RNA, and protein. When I performed my first successful DNA extraction and saw the small white pellet of DNA at the bottom of the tiny tube, I felt wonder at being able to see something that had previously been only a theoretical concept from my classes. The rush of validation I felt every time one of my experiments was successful helped me love molecular biology, and I worked in research labs throughout my undergraduate degree.

I loved filling my lab notebook with incremental results until it was full and I could crack open a brand new one. I loved writing my initials and the date on colorful lab tape and filling boxes with tubes full of DNA before neatly filing them away in the freezer. I loved coming into the lab in the morning and taking agar plates out of their warm growth chamber to find tiny colonies of bacteria growing, showing me that my experiment had worked.

When I was healthy, I could stand at my lab bench for hours, working without rest. I could navigate through loud and crowded lab environments and easily avoid obstructions that blocked walkways. I could carry heavy materials and work with dangerous chemicals without fear. I planned out experiments knowing that I would be healthy enough to do intricate and intensive work in the coming days. However, when I became disabled, the amount of energy that I had to devote to maneuvering through lab environments multiplied tenfold.

I became disabled while a senior in college, two months before I applied to molecular biology graduate programs. I suffered a traumatic brain injury that, although diagnosed as mild at first, resulted in a multitude of symptoms including frequent migraines, hypersensitivity to noise and light, nausea, dizziness, severe fatigue, memory problems, and emotional dysregulation. My doctor's initial prediction was that I would be better within a few days or weeks at most. But my doctor hadn't recognized the severity of my symptoms and grossly underestimated the timeline for healing.

When I first started my PhD program ten months after my brain injury, I did not consider myself disabled because of everyone's insistence that my symptoms were only temporary. For nearly a year, I was unaware that a traumatic brain injury could be considered a disability. However, I apprehensively made an appointment with Student Accessibility Services (SAS) because I knew that certain program policies were going to be difficult to comply with.

I learned during orientation that my program had enacted an ableist policy mandating attendance at every session of a year-long course that met between once and multiple times every week. Only one absence was allowed; any additional absence could force me to retake the course the following year. This policy discriminated against disabled students by refusing to acknowledge that students with chronic medical conditions need flexibility in attendance beyond one day per year when they are allowed to be sick. Because of my chronic fatigue and migraines, I knew I could not commit to such an unreasonable attendance policy. Moreover, some meetings of the course were inaccessible because we were required to attend loud and crowded poster sessions that elicited painful and disorienting symptoms. Since my injury, high ambient noise levels inhibit my ability to understand spoken dialogue and trigger migraines that make it incredibly difficult for me to function for hours or even days because of the pain and nausea. Given my inability to participate in these meetings in a meaningful way and their detrimental effect on my health, I sought an alternative solution that could allow me to fulfill this requirement.

When I arrived at my initial appointment with SAS, I brought all of my relevant medical records as requested, and the SAS employee immediately put them aside, stating that they did not care what the records said because they wanted to hear what I had to say. While I wondered why I was made to go through the effort of retrieving my medical records if they were irrelevant, I ignored the sour feeling and proceeded to explain one of the major reasons that I was there.

Instead of offering flexibility, understanding, and the willingness to advocate on my behalf, the employee suggested earplugs and denied my request: there would be no exceptions to the mandatory attendance policy. I sat there dejected and invalidated as the employee listed the myriad values of attendance, as if I didn't know that it would be a benefit to be able to attend. After I left this meeting, I did not reach out to SAS again until after this employee had left.

As I had predicted, I did need to miss multiple sessions of the mandatory year-long course during my first year. In light of my previous experience with SAS, I had no one to reach out to when one of the program coordinators insisted that I get a note proving that I was too unwell to attend class. The coordinator threatened that without a note my absence would be a "big issue." While dealing with intense dizziness, fatigue, and pain, which had prohibited me from attending class, I had to travel to the student health center. The providers were bewildered by my program's request for proof of illness because they could not reveal any pertinent medical information because of privacy laws. After that experience, I forced myself to attend class sessions even when I had horrible migraines and was nauseous to the point of throwing up in the bathroom during breaks.

While I pushed my body to conform to inaccessible program requirements, microaggressions from classmates, professors, and administrators undermined any remaining joy and satisfaction I felt about science. In one instance, when I was about to give a presentation in a graduate student seminar, the professor loudly said, "You realize you should be standing up, right?" I was exhausted and dizzy and had been sitting while I loaded my slides. This blatant admonishment in front of my classmates created an uncomfortable situation in which I could either disclose my disability to everyone or stand up for the duration of the presentation. I chose not to call out this ableist behavior because it would take more energy to argue than to push through the presentation while standing.

As I continued in my program, I suffered another traumatic brain injury. Again, its severity was underdiagnosed. With my symptoms exacerbated, I could no longer hide my disability and decided to disclose it to several colleagues and supervisors. Although most people were sympathetic, they almost exclusively focused on an imagined future when I would be better. In conversations about my health, there was no room for the possibility of my disabled body thriving in and contributing to academia. Despite the superficial sympathy, I feared that any additional accommodation requests would once again be denied because they would be deemed as compromising an essential element of the program. I had to juggle my second physical recovery with a long to-do list, asking to be excused from seminars and classes with inflexible attendance policies. The syllabus for one of my mandatory classes stated that we were only allowed to miss sessions for "valid scientific activities that can't be planned at other times, such as conferences." While choosing to travel to a conference was an acceptable reason for missing a class, unforeseen illness was not.

During this second recovery, I was finally able to come to terms with the idea that I might never fully retrieve the health I had before. Accepting my body as it was came after a long process of fear, grief, and anxiety, but when I finally let go of the idea of "true recovery," I was no longer spending a significant portion of my energy grasping for an unattainable dream and punishing myself when I could not achieve it. Previously, I had risked my physical and mental health to reach an arbitrary standard unrepresentative of my ability or skill. I went to inaccessible departmental events to be a good "team player," and I pushed myself past exhaustion to finish mountains of work with illogical deadlines. Now, in deciding to reject the pressure to transform myself into a healthy ideal, I gained enough energy to change the environments and policies that were harming me.

After I became vocal about my disability and advocated for my needs, I was disheartened to see how quickly the administrators and faculty went from viewing me as a hardworking, motivated researcher to questioning everything from my passion for science to my ability to manage time to my capacity to work as a member of a team. Once they lost this trust in my abilities, the resistance I met when requesting any accommodation became even more insurmountable than before. My performance reviews worsened, and several faculty members even described how I should simply work harder to "overcome my limitations."

Every tiny matter became about my disability. When I asked one faculty member a logistical question about program requirements, I was told that the program could not "lower the bar" for me. Instead of answering a simple question applicable to all graduate students in the program, this faculty member chose to add my disability into the equation and reveal that disability was understood to be a deficiency. I felt completely unsupported by the people who were supposed to be helping me succeed. My legitimate concerns were frequently framed as excessive anxiety not grounded in reality and were laughed off with the refrain "Don't worry, every PhD student feels like this sometimes."

I started having panic attacks and waking up in the middle of the night, sweating and with my heart racing, replaying every awful comment from students and faculty members. I felt alone and also felt that I could not share how deeply the ableism in my workplace was affecting me without being invalidated or blamed for my own harassment. Just when I thought things couldn't get worse, my department moved to a newly constructed research building that was even more inaccessible to me.

Every day I woke up and mentally prepared for another painful and disorienting day of working in the lab. The crowded layout, bright fluorescent lights, and noise level triggered daily migraines so painful and exhausting that when I returned home from work all I could do was lie in bed while searing hot claws dug into my brain. Throbbing, pulsing, dull aches and searing pain—each day brought a new and painful experience, and I felt that I was doing a PhD while my head was being crushed inside an iron vise.

When I met with SAS about this building, they made it clear that they were unable to change lab environments and could only suggest how individual students could modify their behavior. Attempts to meet with my advisor or other faculty members were similarly unhelpful, and I endured comments disparaging my work ethic when I outlined the inaccessibility of my new workspace. Faculty members suggested I take a leave of absence until I was no longer disabled or else withdraw permanently. When I pointed out that, despite all of the barriers I was facing, I had remained productive and had consistently met all of my research goals (with plenty of evidence to support my claims), the faculty still didn't believe me. They said that it would be impossible to work in the lab for only thirty-five hours a week and still make sufficient progress.

They didn't see that, before I went into the lab, I planned every minute of the day so that I didn't waste time while I was there. They didn't see that I did all of my analyses and any work that could be performed remotely at home so that I could minimize the amount of time spent in migraine-inducing conditions. My planner was packed with notes and schedules so that I could often simultaneously complete multiple experiments instead of performing them one after another. The efficiency that I have honed to maximize my productivity in an inaccessible environment is impressive and speaks to disabled people's resiliency under antagonistic conditions. Unfortunately, when all that is valued is the number of hours one spends in the lab, the amazing strategies that disabled researchers have developed to work in hostile environments (and all of the work that we do while we are not physically in the lab) often go unrecognized and devalued.

I deserve to be in science, but I don't know if science deserves to have me. I get so angry sometimes that I've worked this hard to be where I am but that people who know nothing about what I go through get to decide what I'm worth. Over the years, the little moments of unchecked discrimination finally overflowed to show me that I am not going to be welcomed as a faculty member in academia. I've been treated horribly by many people who claim they have good intentions while they speak over me explaining how they could help. As they tell me that I won't succeed with my disability, that I don't have enough motivation, that I don't work as hard as I should, I realize that I don't deserve to be treated like this. Nothing they say is true.

I'm not lazy for taking the elevator instead of the stairs. I'm not unsociable because I refuse to attend inaccessible events. And I'm not unmotivated because I take time off to rest and recover from an inaccessible environment that the administration refuses to change. I should not have to display superhuman efficiency and resilience just to be allowed in the room. But I can't risk my mental and physical well-being while I wait for academia to decide that I'm worthwhile to keep around.

I grieved when I finally realized that I would not be able to pursue the scientific career for which I had spent years preparing. I felt that I had made all of the right choices and still was being pushed out. Starting in the fourth year of my PhD, I cried myself to sleep most nights at the hopelessness of my situation. I remember turning to my fiancée and whispering, "I wish I never came here. I wish I had never applied to graduate school. I've never felt as bad about myself as they've made me feel here."

I hated that something I used to love would now be associated with relentless pain.

My grief encompassed the loss of not only my dream and ability but also of my idealistic view of science, as well as the fact that nothing seemed within my control. However, while these losses were painful and traumatic, I eventually began the process of acceptance and started to imagine a new scientific future. I now envision a new scientific community that proudly values all people working within the field and the people affected by the research as much as it currently values prestige and funding. In this future, I will no longer have to fight for access to scientific spaces and be treated as a burden. And by prioritizing access and equity, I hope that one day everyone will gain what I have gained—the realization that while we love what we do, we are more than our work, and no environment that exploits and mistreats others will ever be worth the labor it produces.

Leehi Yona

24

(IN)VISIBLY ERODING BONES, BODIES, AND LANDSCAPES

LEEHI YONA

Content warning: surgery

'Ve come to accept that being a climate scientist means being in a lot of physical and emotional pain.

As a climate scientist, I face the injustice and global suffering of this crisis every day in my work. I oscillate between fiery hope and deep despair and grief. Some days I awake undeterred in my faith that a better world is on its way. Other days an insurmountable dread pins me to my bed.

As a person with hypermobile joints and osteoporosis, I also feel physical pain on a weekly, if not daily, basis. Fieldwork is a dilemma: at once awe-inspiring, exciting, and invigorating and also painful.

At times both the world and my body seem to be eroding in different ways. I can't distinguish one pain from the other. And both are, for the most part, invisible to the outside world.

When I was a first-year student in college, I was offered an incredible opportunity to be a field assistant for a graduate student conducting research in Kangerlussuaq, Greenland. I say incredible because I had been rejected from every summer program I had applied to that year. But on the last day of a class on polar science and policy, my professor asked me if I'd be keen on leaving for the Arctic in a few weeks. I enthusiastically said yes—and then realized what I'd gotten myself into.

As a first-generation college student, I didn't own hiking boots. To be honest, I didn't even know what hiking boots were and why they were different from running shoes. I didn't know how to "hike." But I was given a budget to reimburse the cost of any essential supplies. To this day, I thank that professor, Ross Virginia, for giving me that budget—I wouldn't have been able to afford it otherwise.

I bought hiking boots and a steel frame backpack. To break the shoes in—something I thought people only had to do for high heels—I went on my first hike.

I was very nervous. I spent the next few weeks trying to run (in running shoes) outside my house, hoping to build the stamina to be a helpful assistant. I was not the most fit person in the world, in part because of my joint pain. I didn't want to let my graduate student or Ross down.

When I finally arrived in Greenland, I relished the views of the glacial lake by our campsite, right next to the ice sheet. I had never seen anything like it: the ice sheet was the purest shade of white I have ever pictured. I remember thinking that, if I were to paint the landscape before me, I'd use white paint straight out of the tube. The graduate student I worked with, Julia, was the best mentor I could have asked for, an eternal optimist, a wise source of advice about grad school, and a patient teacher. I think she knew that my body was under considerable stress already because I was usually slowing her down.

At the end of every day working in the tundra, I would remove my socks, peel off the athletic tape holding my feet stable, and rest my legs in the icy water. By the end of the field season, my skin was raw from the daily ritual of taping my ankles like a corkscrew. Tundra is notoriously hummocky and uneven, a result of the constant freezing and thawing of the Earth. I rolled my ankles constantly, usually every hour, but the tape helped me make it through the summer. At the time, I wasn't yet aware of my medical condition; I only knew that the tape reduced the pain.

Watching the Greenland ice sheet calve off and break was an indelible part of that summer. Thundering sounds around dinnertime were always a good sign that it would be a big night. The best time to see the glacier was around midnight. Other field assistants and I took our sleeping bags and hot tea (or whisky) and hiked up the mountain ridge just above camp, where we could see the edge of the ice sheet. The setting sun glowed, reflecting gold against the bare mountain ridges. At the top, we sat facing the ice sheet and waited. We heard it first—a loud, unmistakable thunder—and then quickly scanned the face of the ice sheet for movement. At once, a chunk of ice the size of five skyscrapers broke off the glacier and crashed into the water below. Perhaps this is one of the few times the emotional pain of climate change felt physically, viscerally visible. I have never felt as insignificant as when faced with the weight of those icebergs.

Not long after the Greenland field season, I discovered that I had hypermobile joints and, at twenty-two, osteoporotic bones. (Every time I get a bone density scan, no matter how many years apart, the same radiology technician says, "A scan? At your age?") My hypermobility makes it more likely that I will trip and fall; my osteoporosis makes it more likely that I will break a bone if I do. My usual approach to fieldwork is to tape my ankles until the skin is raw to keep them from collapsing into each other. If you went into the field with me, maybe you'd notice that I am consistently last on hikes and field courses, but the tape is usually hidden in my socks. I am able to walk and hike and do pretty much anything, perhaps just more carefully (and slowly). It's not just my ankles: in graduate school I went to an occupational therapist who first introduced me to finger splints. The first time I wore them, I realized that pain was indeed my status quo.

I hadn't noticed that my fingers were hyperextending as I typed; I learned to dictate.

My pain is often invisible: you would never guess that I have any medical conditions if you met me. This invisibility isn't always a good thing, especially when combined with a high tolerance for pain.

While on a policy research trip in May 2017, I tripped on a recreational hike, but the pain wasn't bad enough to suspect that anything was wrong. As a matter of fact, I had begun training for a marathon that same month, and by July I had completed a half-marathon training run. I wonder why I ignored the pain. I think I just wanted to prove that I was finally capable of being active, and I was embarrassed to see a doctor, suspecting I'd be told that the pain was nothing, perhaps worried I would be judged for how I looked. That month, I found out that I had fractured my tibia. The same day that the doctor called to give me my MRI results, my friends shared a headline on Facebook: a massive chunk of the Antarctic ice sheet—Larsen C, estimated to weigh a trillion tons—had broken off. Just as my pain had been accumulating for months, Larsen C had been melting for decades. Both fractures were not the result of a day's work but long in the making. My only solace was that my tears were multitasking.

Over two years later, in October 2019, I was in Yellowknife, Northwest Territories, for an academic conference. Everyone around me wanted to go hiking, but instead I walked around at night, mesmerized by the auroras. My stomach hurt. I attributed the pain to the temperature change and dryness of the Arctic, to something I had eaten, or to exhaustion. It is quite easy to ignore your own pain if you dismiss it long enough. I should have realized that something was wrong when I didn't have enough appetite to eat for two days.

By the time I was flying back to San Francisco, I was worried. The pain was on my right side. Before my connecting flight in Calgary, I asked the gate agent if my second flight was far away. When she said it was a fifteen-minute walk, I asked if I could use a wheelchair. I felt awful, a healthy-looking twenty-something being pushed by a woman who had been working in the airport since the 1970s. But my backpack was heavy with textbooks, and I finally entertained the thought that maybe I was in enough pain to be experiencing a medical emergency. I landed in San Francisco and went straight to the hospital. Not only did I have appendicitis—my appendix had already ruptured.

When the doctors came in and told me the news, I emailed my PhD advisors with the subject line "Burst Appendix," apologizing profusely for needing to miss any meetings that week. I can't recall this moment without cringing, incredulous at my past self for being more concerned with research than my health. I still don't know what I would have done differently, since graduate students are often expected to make research their top priority.

I am lucky that my advisors were supportive: they checked in on me and insisted on visiting me in the hospital. After my surgery on October 31 (it was a very scary Halloween), they called me to make sure I was okay. They told me to prioritize my health above my research. I hadn't expected this because I thought that academia's culture of overwork and of prioritizing research would prevail above well-being. I am grateful to my advisors for supporting me and helping me realize that getting through the semester healthy was in and of itself a worthy goal.

Recovery took longer than expected, and I was back in the hospital a few more times over the next month with a lingering abdominal infection. In December I was scheduled to present a poster at the American Geophysical Union annual meeting, my first major milestone for my PhD. I didn't have the results I planned to gather by then. All of my advisors reassured me that if I didn't want to, I didn't have to present a poster. One of them said, "Just put up a piece of paper on your poster and write, 'burst appendix.' " An immense weight was lifted off my shoulders.

I decided to go anyway, mostly because I am stubborn (or, as my friend Hannah says, "tenacious"). I didn't want to let my advisors down. Just attending the conference for this one presentation was a success in and of itself. I wore my loosest pants so that they wouldn't press against my surgical scars. I did my best to present my poster without vomiting (at which, thankfully, I succeeded).

When illness is acute, people notice. People help. When a massive wildfire hits the Pacific Northwest or a hurricane ravages the Atlantic, fundraisers spring up. People keep impacted communities in their thoughts. But when coastal flooding leads to the gradual erosion of a landscape or fossil fuel extraction slowly poisons low-income communities of color, we are less likely to notice. A burst appendix demands to be noticed. Chronic conditions are a hum in the background, a pain that is always ignored.

After my appendicitis, I was bedridden for nearly a month. Alone in my studio apartment, I was, quite simply, sad. I told my family not to visit, arguing that I had friends to help me. Indeed, I did have the most supportive friends. But the real reason I told them not to come was that I was planning to take the LSAT later in the month, a spontaneous decision to apply to a joint degree program. But I knew immediately that telling my family not to visit was a bad decision. Once again, I had prioritized school over my well-being, and I was miserable.

I told myself that if there ever was another such decision in my life, I would choose family and my health. So when the coronavirus pandemic began its tidal wave around the globe, I didn't hesitate: I booked the next flight out to Montreal to be with my family. It was not the best decision I could have made for my productivity or professional success, but it was by far the best decision for my health and overall well-being. And, for once, I am proud of myself for prioritizing my well-being.

As an interdisciplinary climate scientist with a chronic condition, I have to strike a delicate balance between challenging myself and knowing when to say no and rest to work another day. My body is its own ecosystem that deserves care; if I neglect it, it's only a matter of time until my body, like our planet, will demand to be heard.

Person drifting in the sea

25

THE ABYSS

KATIE HARAZIN

Content warning: alcohol use, suicidal ideation, mention of blood

"**M**a'am, you're injured—are you okay?" asked a strange voice. I murmured some response, I think. I didn't notice the pain I must have felt. In the early morning hours of New Year's Day, I lay limply on the sidewalk. My hands rested against my face, and a trickle of blood tickled my ear. I was hurt and alone.

The memories are like stills cut from a medical thriller TV show. The EMTs were helping me into an ambulance. I saw my face covered in red as I glanced in a mirror. There was the indistinct chatter of busy nurses in an empty lobby and the shrill echoes of phones ringing down the hallway.

I don't remember my face crashing into the curb because of the alcohol sloshing around in my brain, but I recall the events leading up to it. I remember that I had polished off a bottle of pinot noir by myself before sunset. I remember skipping out of my home to celebrate with friends. I remember drinking beer and taking shots. I remember I didn't stop.

I wish I could claim this was a fluke, a single event attributable to a party. I wish I could spare myself from the humiliating reality.

For years, I denied the disturbing trend in my behavior and the consistent intensification of problems. I'd woken up to mysteriously cracked phone screens. Friends told me about arguing with strangers the night before, and I'd received responses to texts or emails I swore I never sent. Once I even found myself at the police station and, of course, coming to my senses in the emergency room, my face all bloody from a fall I don't remember. My life was spiraling out of control as I increasingly turned to substance abuse. It still managed to catch me off guard.

The truth is: I did notice the pain, but not from the road burn on my palms or the cut in my face. This pain was one that ceaselessly gnawed from within, a wound in my soul that I desperately tried to heal.

This self-destructive creature was a mere ghost of the person I had been before my PhD. Three years before, I had been allowed into a PhD program in Australia to pursue a degree in Earth and ocean sciences. It was a dream come true, and I was elated. "This is my golden ticket," I repeated to myself. I uprooted my life and journeyed ten thousand kilometers from home to study the world beneath the deep and wide abyss of the ocean.

Below the playful dolphins and bright coral reefs, darkness cloaks the bottom of the sea. Light can only penetrate so deep, and the overlying water creates a bone-crushing force. This suffocating space reaches near-freezing temperatures, and the abyssal plains extend flat and featureless forever. Here terrifying creatures adapt and thrive in a treacherous environment. I was captivated.

My first eight months were exhilarating as I discovered a whole new continent and made new friends. As the days passed, the excitement became overwhelming to the point of exhaustion. The usual hurdles of the marathon that is a PhD also began to affect me, multiplied in intensity by relationship heartache and homesickness. In this period, a friend of mine died in a sudden accident, adding to my grief. Nevertheless, life demanded longanimity, and I yielded. I worked diligently and earnestly participated in meetings. I even produced scientific research results. For months, I strained myself to remain optimistic and confident. Despite my efforts, however, I always seemed to come up short.

The PhD abroad wore me down. The long hours of tedious work and the frustration with interpreting my results made me question my competency. Surely, I didn't have imposter syndrome: I didn't just think I was a fraud; I was convinced that I was a fraud. I felt left behind as my cohort charged ahead with their progress while my research seemed to stagnate. The work made me anxious, and the anxiety made work more difficult. As the stress grew, so did my need for stress relief. Alcohol emerged as the distraction.

I kept falling behind in my work, and few took notice—except that my advisor demanded doctors' notes for absences (for going to therapy) as if I were in sixth grade. The student-advisor relationship soured.

One day I gave a presentation that I had been told would be an informal discussion with my small lab group. To my surprise, others joined, including a high-profile guest from another institution, filling more and more chairs. As I stumbled through the presentation, thinking out loud about how to improve my approach and what I needed to do, a committee member interrupted me.

"If this is the highest-quality work you can deliver, you might as well be a trained lab monkey," he opined ten minutes into his rant. He only stopped after I told him to.

From then on, the memory of this colossal humiliation made simple presentations a source of panic attacks, so I desperately avoided them. After that, I refused any offers to present. My first attempt to give a talk to fellow students many months later resulted in me hyperventilating in the bathroom.

These experiences exacerbated my underlying mental health issues, including chronic depression and anxiety.

I am not sure exactly what caused my breaking point. Did I hit a wall for my little lab monkey brain? Did I lose my purpose and direction dealing with grief from the sudden death of a friend or a failed relationship? Many things collided at once to reignite these issues. I suffered from insecurities, persistent mental fog, self-loathing, sleepless nights, and a memory bank constantly replaying every embarrassing moment in my life. There was a perpetual shroud of pressure and uncertainty. It was my personal abyss.

I marked the start of my fourth year as a PhD candidate with my inebriated encounter with the pavement. The stress of my work overwhelmed me as the tension with my advisor increased. I wasn't coping well, to say the least. Maybe I could have left the lab group, but my other options seemed limited. Importantly, my attachment to my research was too strong. I had found an intense passion and had created a raison d'être around it. I had formed a fantastic dream of being a scientist in my head and pursued it endlessly, with great emotional and financial investment. I scrambled to regain my posture. Whenever I presented a product, I was met with disappointment. Where I had hoped to receive advice when I was struggling, I instead received cruelty. The isolation I felt expanded with every scathing email and meeting.

I was hurt, and I was alone.

I used alcohol to cope. I was too drunk or too hung over to function or too foggy between binges to work. People in my life saw these consequences, but most—including my former advisor—brushed off my truancy and avoidance of interaction as "laziness," "disinterest," or "not being cut out for the program." On top of that, one of the hard problems with mental illness is that the incessant, intrusive thoughts and habits become a reinforcing, vicious cycle. Untamed depression entices a drink, the drink exacerbates the predicament, and the escalation sometimes brings pills and powders into the picture. This situation further increases the depression, and the circle repeats. When I gaze at that abyss, it gazes back.

At this point, I was fully entangled in my downward spiral. Weirdly, though, there was an inexplicable aspect of control— one that felt essential when everything else was spinning chaotically. Drinking vodka at ten A.M. seemed unhinged, but when I woke up seriously evaluating the costs and benefits of staying alive, it was a panacea. It staved off the immediate threats of the abyss, yet it also scratched that itch seeking long-term self-destruction.

I was still months and months behind on my milestones, but I had good data and inadequate drafts. Despite my fragmented path, my advisor told me I had everything I needed to finish. I knew I had all the tools and that I could succeed. But things fell apart. One day my committee decided to meet, leaving me in the dark. In this meeting, they decided to terminate me without giving me any chance to fix things. I was given few options of where to go next but plenty of reasons why this decision was immutable. In other words, I received a three-page letter describing how I was "stupid" and "lazy," with no path forward except being thrown to the wayside.

I was ashamed beyond all measure, and my aspirations and dreams to be a scientist crumbled before me. I was devastated.

I had finally made it all the way down, and I could see the bottom of the abyss.

An abyss, by definition, is forever expansive and unending. There is no bottom. In contrast, suicidal ideation imagines a boundary to and an escape from the abyss. What I saw was a dire solution to a dire problem. It was terrifying. I desperately avoided the bottle. I didn't feel safe by myself. I asked to stay with a close friend for a week because I didn't—and couldn't— be alone.

At the same time, almost paradoxically, a survival mode kicked in. I immediately consulted with the dean of students and other people in the college who might empathize or understand my situation, who might perhaps see others in my situation, or who even might have experienced a similar dilemma. I furiously typed a rebuttal to argue that I was still worthwhile—partly to maintain my enrollment but mostly to convince myself that the culmination of my life's efforts was not for nothing.

During this rare period of sobriety and determination, my efforts triumphed. The gambit worked. I could stay.

This result did leave me, however, with an uphill battle of completing my work independently while assiduously managing my mental illnesses and their detrimental effects. My advisor shunned me, even when I swallowed my resentment and pain to ask him for help. I worked alone to meet a mountain of goals. The cumulative damage of the previous years had an enduring impact on my well-being.

Hopefully, I'm officially done with my PhD by the time you're reading this. As I write this, my thesis document taunts me from the background, a minimized window on my desktop, so close to completion. Even if I'm not done, I will be eventually. I'm not pressed. The important news is that I'm doing better now.

My "rehabilitation" has included moving back home to the south of the United States to live with my family. Here I have resources and healthy distractions. I have more space to move around, and I have a dog that gives me affection and joy. I joined a local lab group that supports me and all its members, and I found a new science community that I thrive in. I've pursued art, a pastime long loved yet long neglected. I am equipped with the mental coping tools that amazing therapists have given me. I'm working, supported, at my own pace: slowly perhaps, but constantly moving forward.

Not everyone has these privileges, and I recognize that I'm a lucky case. I've been able to maintain my PhD program, although I was only partly functioning. I've had avenues of financial support. I am not fully consumed by substance abuse anymore. I have not gone into inpatient recovery, either willingly or unwillingly. I have not been arrested or incarcerated unrightfully because of my mental health. I have not been taken to the ER for an overdose or an attempt on my life. Most importantly: I am not dead. I can see the hope beyond my own abyss.

Sunshine Menezes

26

NAVIGATING THE CURVE

SUNSHINE MENEZES

Content warning: cancer

R ecently, social media—an unrelenting recorder of memories—reminded me of what I looked like four years ago: slim, bald, tired, relieved. That photo captured my last day of chemotherapy to treat a breast cancer that had metastasized from a couple of tiny, precancerous cells to a liver filled with tumors, bulging like a sack of marbles.

I've been living with cancer (two cancers, actually) on and off again for ten years now, but the language of this experience is hard to capture.

As a scientist, precision is important to me. Language is important to me. The imprecision and tortured language of chronic illness, and specifically cancer, is something that I struggle with on a regular basis. I have learned that medical treatment, at least in some cases, is as much art as it is science. I've discovered that I don't want to be described as "brave," a "warrior," or a "superhero" just for playing the cards I've been dealt. I don't want people to dwell on my "fight."

Nonetheless, it is true that I live in a body that is permanently changed. This experience has forced me to consider the complicated ways we construct and navigate our self-perceptions. Why, for example, do I find it so difficult to accept the idea of being chronically ill?

Much of this struggle relates to the whiplash that comes from passing through distinctly different states of wellness. It feels as though I had been driving along a straight, flat road that was well marked. Maybe I drove too fast sometimes or got a little sleepy and hit the rumble strips, but the road remained easy to navigate. Then, suddenly, a sharp curve appeared in front of me. I had to grip the wheel tightly and carefully balance the required amount of acceleration and braking to escape a deadly crash. The physical and emotional toll of that curve was exhausting and sustained.

Then, over some years, the road straightened out again. My car had some serious, lasting damage, but it still worked most of the time . . . until I hit another one of those unexpected hairpin curves and had to deal with the same thing all over again. But with this second curve, a second and metastatic cancer, the situation changed permanently: I'm now driving a damaged car along an endless curve.

Living with metastatic cancer puts me in a strange sort of limbo. On the one hand, I'm well right now, in remission. On the other hand, there is a quietly ticking time bomb in my trunk with an unreadable clock. Will it detonate just as the passing landscape dulls my alertness?

This uncertainty is another aspect of my struggle to view myself as chronically ill. When driving along a sustained, gentle curve, it's possible to occasionally forget that the curve is there. When I remember the curve, though, it can be terrifying, completely consuming my thoughts. I'm so attuned to the subtle shifts of my body that every irregularity can make me fear the reawakening of dormant cancer cells—a fear that only subsides with my next blood test or CT scan. I haven't relied on religion or meditation or self-help books to cope with that terror. Instead, I've applied the hyperrational and critical thinking skills I developed as a scientist to learn all I could about my two cancers and their various treatments, focusing on the facts and using the data to make decisions. The process of earning a PhD in biological oceanography taught me how to do literature searches, read and interpret scientific journal articles, and identify important questions. Specifically, as a taxonomist, I studied how organisms are classified into different groups. In other words, I was trained to be particularly attentive to the smallest details. While I may have to look up every fourth word in the medical literature, I have a pretty great bullshit detection meter that has helped me do a reality check on the vague reassurances of well-meaning, but insufficiently direct, doctors. This bullshit detector also has helped me advocate for the validity of my experiences as a patient, even when the doctors insist that those experiences are unlikely.

With each of my diagnoses, my scientific training kicked into action and I focused more on research. There was no paper too tangential, no Internet wormhole too deep. This focus on evidence was also part of my emotional well-being; I needed something certain in the midst of so many unknowns. I gathered my small army of scientist friends, each with their own special research skill set, to help make sense of the studies I couldn't understand and to help me wrestle with treatment decisions.

When I discovered that the meta-analyses related to my first curve in the road, uterine leiomyosarcoma, were based on fewer than three hundred cases that mostly included women much older than I was, I immediately recognized that I was in uncharted, or at least poorly charted, territory. A few years later, when I hit the second curve of HER2+ breast cancer, I was relieved to discover that this subtype was one of the best-studied cancers, providing a wealth of evidence to inform my decision making.

Still, my scientific training didn't provide the tools to manage existential questions of life, death, or limbo. During my most intensive periods of treatment, I chose to lock up the very complicated, often conflicting feelings of the experience because it was easier than being consumed by them. Since then, I have often been angry, frustrated, disappointed, and heartbroken by the uncertainties of these diseases. I have felt—and often continue to feel—broken. I have felt unknowable. The lingering dread of waiting for the next hairpin curve has become as natural as breathing.

Even while I wrestle with these feelings that are so common among the chronically ill, however, I see the cognitive dissonance playing out in people's eyes when they discover I have metastatic cancer. After all, I look perfectly "normal" now. I'm sure I have countless interactions every day with people who don't know my medical history and who assume that I am completely healthy. I am both annoyed and pleased by those responses: annoyed because the inevitable exclamations of "you don't look like you have cancer!" demonstrate a lack of appreciation for the fact that chronic illness doesn't "look" a particular way; and pleased because this response reflects my relatively good health, at least for the moment.

So, which is it? Am I sick? Am I well? Realistically, I'm somewhere in between, and even with the immense privilege of receiving top-quality medical care, no one can predict where I might fall on this spectrum tomorrow or the next day. My desire for precision is in conflict with my understanding of uncertainty. My desire for clarity is in conflict with my understanding of chronic illness.

My metastatic breast cancer has been in remission for four years now, long enough that I've earned the precious medical label of "exceptional responder" to the maintenance treatment of monoclonal antibody infusions I continue to receive every three weeks. This treatment regimen is new enough that there's insufficient evidence to say whether it's safe for me to stop. The only way to know would be to take the plunge, and that seems like a ridiculous, unnecessary risk. So I continue to receive my regularly scheduled treatments. So far, they continue to do their job.

Beyond this treatment, living with my chronic illness is mostly about managing the physical and emotional aftermath of four surgical procedures, two extended periods of radiation therapy, and eighteen weeks of chemotherapy: the occasional, out-of-the-blue soreness and discomfort where I had a lumpectomy; the annoying challenge of turning a page when neuropathy acts up in my fingertips, eliminating the typically unnoticed sensations that drive such a small act; the hesitance to spend a day at the beach or go for a hike because of the ever-present potential for a random bout of diarrhea, a side effect of my ongoing treatment; the extreme exhaustion on days when the diarrhea comes in waves; the strange normalcy of menopause at the age of forty-eight, ten years after I was suddenly thrust into a new stage of life when I was diagnosed; and, certainly not least important, the perpetual sadness that I was robbed of the chance to experience pregnancy.

I have gone back and forth between denial about my chronic illness and what I can only call intense nondenial. "Acceptance" isn't quite right. It's more like a recognition that the road is unlikely to totally straighten out and could take a hard left at any time. I sometimes wonder if my driving skills are still good enough to handle that next surprise turn.

Perhaps, then, my struggle with my identity as a chronically ill person is deeply connected with my identity—and thinking—as a scientist. Perhaps I'm so accustomed to being attentive to minutiae—to the painstaking process of observing, reading, experimenting, and adapting—that the day-to-day *regularity* of living with cancer makes me uncomfortable with the language available to describe my reality.

Some days I feel like a chronically ill person. Some days I don't. So I continue to straddle the centerline of this road, trying to enjoy the scenery while I can.

Olivia Bernard

27

TIDY COLUMNS

OLIVIA BERNARD

Content warning: medical testing

Once a month the Earth, moon, and sun align, causing the tide to rise higher and reach toward the trees that border the Great Marsh of Massachusetts. The water fills the creeks and creeps over the tall grass, flooding the dry land. Sea critters weave between the Spartina blades in a temporary utopia. And scientists head out onto the marsh with their waterproof notebooks and neoprene boots, combing through the grass to count the crabs and eels. Their hands smear mud down the page as they record their data in tidy columns.

It might have been a pickup truck, but the Ford Ranger felt small when three people were sitting up front, and my short legs didn't help. I reached under the seat and pulled the bench forward, taking my two passengers along with me. But it was only a short ride to the field site, and as I drove I began to worry. I repeated in my head what my supervisor had told me: just be a penguin. I visualized my arms loosely by my sides as I pushed off, leaned away from the creek's edge, and glided across. But it was two in the morning, the water would be dark, and I wasn't a very good swimmer.

I parked the truck next to a narrow path that would take us to the salt marsh. It might have been a hot summer, but the nights were still cold. We took off our sweatshirts so that we could come back to something warm and dry. As we walked through the woods, I thought about being a penguin. I also thought about being cold and itchy. I thought about why I was so itchy and looked around for poison ivy, but it was too dark to see. I pulled at the collar of my shirt and rubbed my hands together as my palms started to tingle and then burn. I held my hands up to my face, searching for any signs of scrapes or bites. The burning crept up my arms and neck and over my cheeks until it reached my eyes. I pressed my hands against my face, trying to ease the pain. I stumbled forward, over roots and rocks, until my foot plunged into cold water. My spine stiffened and my mind became quiet as a chill shot up through my bones. I stayed like this for a moment, not stuck but grounded, and then opened my eyes.

Water rippled around my ankle, a vibration sent out into the universe. The water was dark and still and stippled with stars, and it went on in all directions. I stepped away from the trees and moved through the water and into the marsh, looking down at what was dry land only a few hours before. The tall grass now only peeked past the water's surface, as if reaching from one world into the next. We neared the salt pannes, and with one false step my leg was swallowed by the mud. I grasped at the ground around me until I found a plank and pushed on it with my hands until I could raise my leg out of the water and mud. I stood on the plank, teetering on this relic of a familiar world as we ventured deeper into the dark, moonless expanse. We followed a soft arc of stars until our nets came into view. As I neared the creek edge, I let my arms hang loosely by my sides, pushed off, leaned forward into the water, and softly swam through the night sky.

Like any good scientist, I kept detailed records. I didn't track every symptom because some came and went infrequently, but others never stopped. The itching, which quickly escalated to hives and swelling, earned a row in my spreadsheet. I wasn't just keeping records, though; I was performing experiments. After a few cold nights, I spotted a pattern in my data and formed a hypothesis: if I hold an ice cube to my arm, my skin will swell. I got what I wanted, a positive result, but the causes of other symptoms remained elusive.

At home I made a new row in my spreadsheet and colored the cells green. This column was for a familiar fog that overtook vision in my right eye as I drove home. While I'd had this symptom before, I still followed my procedures, trying to collect data. Did it go away when I rubbed my eye or blinked? Did I have a headache? What was the pattern? As I added it to my spreadsheet, I remembered the vision problems I had as a teenager. As I pulled binders and notebooks out from my closet, loose pages and Post-It notes fell to the floor. My answer would be in the box of unorganized records thrown together during a time when my symptoms were strange but nothing more. I thumbed past the test results, the list of symptoms and diagnoses, to my appointment calendar. There it was, a visit to the eye doctor and then the neurologist. I searched the box for hints of what I was doing those days, what I was eating and how long I was sleeping. I laid out the evidence across the floor, desperate to understand.

"Do you have any kidney problems?"

I sat in the hallway wearing a hospital gown that was much too large and disposable socks that did nothing to keep my cold feet warm. The receptionist sat at the end of the hall scrolling on a phone with a bag at their feet. It was late in the evening. I had been scheduled for a last-minute appointment, an inconvenience. But my records didn't give me answers, and I needed more data.

"Well, yes, but not kidney disease. Why?"

"The contrast agent can cause damage to your kidneys if you have kidney disease," the technician replied as they taped the IV to my hand. I always hated having an IV in my hand.

"I get cold hives; do you think I could react to the temperature of the contrast?" I asked. The technician seemed unsure of how to answer. They motioned for me to follow them into the MRI room but briefly paused in the doorway and told me that they would inject only a small amount and I shouldn't worry.

As I lay on the table, the technician explained that the machine would be loud and the space would be small but that a strategically placed mirror would help ease any claustrophobic feelings. They placed earmuffs on my head and covered my face with a cage, and then they gave me a remote, a panic button, in case I needed it. I breathed deep into my belly and looked down toward my feet as they pushed me into the machine. As I lost sight of the room, I slowly breathed out and looked forward.

A clown fish gingerly swam between the corals, in and out of my vision. Gentle waves rocked a pebble back and forth on the sand. I held tight to the remote, my tether to a familiar world, and listened for muffled guidance from the technician. Electricity surged through the coils of the machine and sent heavy vibrations throughout my body. I became suspended, staring through the porthole to a world just out of reach. Yet this small space felt vast.

I heard a click in my ears and a voice followed: "If you don't like the aquarium scene, we can change it to something else."

I spread a hand towel on the counter and placed a mug and plate on it. It was a cold night, but I kept the window ajar so that I could listen to the great horned owl, the only other being awake with me at this hour. I grabbed the kettle just before it began to whistle, poured water into my mug, and took my tea and toast upstairs. I liked to sit by the window once it neared four in the morning, looking out into the darkness. I'd imagine setting out to find my lonely owl friend and consider the strange things hiding in the unexplored night. But if I let my mind wander too much, I'd never find it again. So I'd sit and wait for the first set of headlights to weave through the neighborhood and guide me back from my wakeful dream.

I placed my mug, still half full, on the other side of the room. I liked to drink tea when cleaning. When I grew bored, I could take a short break by fetching my drink, but it wasn't so far that I'd get distracted. I opened my closet door and took down the binders and folders and boxes of notes. All of my data piled on the floor: lab imaging, blood tests results, scientific studies related to my conditions—my life's work. One piece of paper caught my attention, a list of my past doctor appointments that started on page one and continued to page seven. Each line included a word or phrase to describe my reason for the visit. The same few symptoms repeated, reflecting a lifetime spent searching.

A few important records made it back into the closet, which was ready to be filled with books and games and other more interesting things. The rest went down to the fire pit in big paper bags. With the lit match in my hand, I hesitated, the reflex of a scientist who was about to destroy their data and forego their discovery. But those tidy columns could not hold my entirety.

They couldn't bring to my memory the taste of still, dark seawater. They didn't reflect the beauty in my life, the humor or the moments of magic. When the match began to burn my fingertips, I dropped it into the fire pit. Maybe I wasn't such a good scientist, and maybe I didn't want to be. I watched as the edges of each page turned brown and then black, until the pages suddenly caught fire and shrank. The fire consumed years of energy in only a moment, leaving me with small glowing pockets of life. And that's what I have, isn't it? In between the flare-ups and the appointments and the uncertainty, I have pockets of life: not to be analyzed but to be acted on and enjoyed. Embers in the ash.

VI

I AM THE CAPTAIN
OF MY SHIP

I am the master of my fate: I am the captain of my soul.
—William Ernest Henley, "Invictus"[1]

On ships the captain has the ultimate authority over all decisions. This section is about authors taking control of their life, career, identity, and plans despite the challenges posed by inaccessibility. It's not about "overcoming" our disabilities (one of our community's pet peeves), but about driving ourselves forward as whole people, including our disabilities. Sometimes that means challenging nondisabled people's misconceptions and stereotypes.

Beginning this section, Jennifer, a young professor, is suddenly paralyzed and must learn how to redefine some aspects of her life. Taylor has lived most of her life by the philosophy "don't be lazy," but when she starts to deal with her depression while in college and graduate school, she embraces a broader view of the world. Gabi, a cave scientist who has worked extraordinarily hard to still participate in fieldwork with chronic pain, finds herself on an actual underground ridge when something shifts inside her. Vincent—a scholar, a Paralympian, problem solver, mentor, and dedicated son—shows us through these different lenses who he is and how he came to terms with becoming blind. Ending this section is Michele, a partially deaf professor who prefers to not wear her hearing aids and challenges the idea that disability is something to be cured or fixed, instead fixating on the strength of her disability.

Many disabled and chronically ill people have developed unique leadership capabilities as a result of the self-advocacy they have to do. No one would say that a captain overcomes the ship—captains work with the ship. We work with our conditions and symptoms and are often stronger because of this work.

NOTE

1. Henley (1849–1903) was an English poet who composed "Invictus" in 1875 while in isolation during one of his life-threatening battles with tuberculosis. Henley was the inspiration for the one-legged pirate Long John Silver in *Treasure Island*, by Robert Louis Stevenson.

Jennifer L. Piatek

28

BROADER IMPACTS

JENNIFER L. PIATEK

Content warning: spinal cord injury, accident

"Think about the opportunities to expand your broader impacts statement . . ." was not exactly the kind of thought I would have expected to be running through my mind at that moment, but in hindsight it was perhaps exactly what I needed to be thinking.

It was the end of my third year as a tenure-track assistant professor. I had just returned from an early career faculty workshop—the kind of event where pretenure scientists like me would listen to presentations by more senior academics about topics such as how to make our scientific research relevant to society (those "broader impacts") and ponder in round-table discussions and "gallery walks" about how we could be successful and happy in our academic careers.

One staple of such workshops is the discussion of a healthy "work-life balance" and the importance of taking time for oneself. So it seemed only appropriate that, instead of driving straight home from the airport, I might take advantage of the lovely June afternoon and do something that made me happy: ride my horse, Jack. I don't know whether it was a stumble or a buck after we cleared the little fence in the middle of the arena (it couldn't have been more than a foot tall, an obstacle we'd cleared without thinking dozens of times), but I went over Jack's shoulder and landed on my back in the sand. The force of the impact caused a burst fracture of my T5 vertebrae and completely severed my spinal cord.

While lying in the sand, my thoughts were not about the lack of feeling in my legs or the fact that, oddly, nothing hurt . . . yet. Instead, I wondered: How would I continue to do science? How would I teach? How could I do fieldwork? Then my "science brain" (as I called it later, blogging about the accident) took charge. It realized that thinking about the negatives—fear, anxiety about suddenly being disabled, worry that suddenly "scientist" couldn't be part of my identity—would be the beginning of a downward spiral into the abyss. My "science brain" responded with thoughts that could act as little flotation devices to keep from sinking into those negative thoughts and fears. I thought about how this injury might not be permanent, how it might be an opportunity to explore new approaches to teaching, and how it might present a new, challenging path.

I fell in love with planetary geology a bit late. I'd spent most of my childhood thinking I'd become a veterinarian, but then I was taken with physics in high school. My physics teacher had made a big deal about the opposition of Mars in 2001 as the best opportunity to send astronauts, and the idea had hooked me. When my college astronomy courses skipped over the solar system entirely, I found a summer job working on solar system research with a geology professor who studied the composition of asteroids using spectroscopy. The idea of using light to remotely detect what something was made of fascinated me, and I realized I'd finally found what I wanted to do with my studies: I'd found my scientific identity.

I was fortunate, in many ways, that this was the path I had chosen: the bulk of my research work is with computers and satellite images, not out in the field where wheelchairs can be unwelcome. Although grad school and a postdoc position had taken me across the country and back, my current university is less than an hour-long drive from where I grew up and where most of my family still lives. The rehab hospital where I spent the rest of the summer after my accident (and where I still go for follow-up appointments) is "just up the road" from campus.

That doesn't mean, though, that I can just return to the same life I had before my accident except with a wheelchair. Many mundane tasks suddenly require new procedures or equipment; getting dressed in the morning is a process (I used to have a spotter to make sure I navigated the shower chair safely), changing a lightbulb requires new tools, and loading the dishwasher involves a complicated dance to maneuver around the open door to reach different parts of the shelves. A decade after my injury, I'm still finding new ways to do old things.

On campus, there are many practical barriers to being a professional who uses a chair that perhaps are not obvious until one is faced with them. Most of our classrooms and labs seem to be designed for only a person who is standing. The computers that operate the classroom projectors, the "lecterns," the whiteboards—they're all best accessed by someone who is five or six feet tall, not someone who can duck under the three-foot barrier in the parking garage. Many labs have tall benches that are nearly impossible to reach from a seated position or are barely wide enough to pass a wheelchair—and that's without anyone else in the room. On my first visit back to campus, I realized that the "handicapped" restroom door was too heavy to open, even if I could manipulate a wheelchair in the tiny stall.

On top of this difficulty is a general feeling that academia is a solo profession. We prefer to have our own offices and lab spaces where we can work on our own projects. Our tenure reviews prefer that we be the first author on grants and papers; our teaching evaluations reflect only our personal performance, even when sharing a course with another faculty member. This attitude is even more endemic in geology, where many of us were taught to work individually in the field, not to rely on a partner for help identifying a rock unit or to locate ourselves on a map. Suddenly I could no longer set up and break down my own lab experiments alone or run a field trip without help.

But the important thing became my identity as a scientist. All the EMTs, nurses, and doctors who took care of me after my accident asked my name and what I did (no doubt part of the protocol for assessing possible head injuries, which, luckily, I did not have—I always rode with a helmet and am thankful for that bit of plastic and foam). I answered all those questions by saying that I taught geology and astronomy ("rocks and stars"); that is what I felt was important. While in rehab, we created a list of "goals" for when I went home: going back to teaching was number one on my list.

I hate to ask for help, but it is more important that I make sure we do our impact crater lab in astronomy class. I get over the discomfort of asking students to lift the ten-to-fifteen-pound trays of sand because I still love to see the different results they get using slingshots and ball bearings to test hypotheses about impacts. I hate the hassle of traveling, particularly being carted on and off an airplane in an aisle chair, but cannot miss the excitement of a scientific conference or the chance to explore a new geologic site.

The culture of academia is slowly changing: there are more opportunities for virtual networking and conferences to avoid the travel hassles, there is more awareness of the need for universal design in our classrooms and labs to maintain accessible and inclusive spaces, and there is more of a sense that collaboration is more productive than isolated scholars toiling alone. I am also fortunate to have found a community of like-minded geoscientists through the International Association for Geoscience Diversity (actually, they found me—it's hard to hide in a wheelchair); when I need resources to help with accessibility in the field or just someone to share similar experiences, I have a group of people to talk with.

There are things that we each can't do and things we can, but they may not always define us. I cannot navigate stairs (safely, anyway), cannot fit a thirty-inch-wide wheelchair through a twenty-eight-inch doorway, and cannot board a plane without assistance. But I can still stare in awe at spacecraft images of Mars, write code to help interpret those images, share my passion for planetary geology to my students and colleagues, cheer for my favorite hockey team, and sing along with the angry music on my "get to work" playlist (as long as no one is listening). In the end, the important thing for me is that I am . . . still am . . . will continue to be . . . a planetary scientist.

Taylor Francisco

29

DOO HWIŁ HÓYÉEDA . . .

A Lesson Lost in Translation . . .

TAYLOR FRANCISCO

Content warning: depression

There's a phrase in Navajo that translates roughly to "don't be lazy." It is usually spelled *Doo hwił hóyéeda*. It's hard for most English speakers to say because Navajo is a breathy, nuanced language, complex and beautiful. I encourage you to search online to listen to a native speaker talking in Navajo. We call ourselves the Diné, which means "people," and our language is sacred. There are so few Navajo speakers living today that I think the words of my ancestors are a treasure every time they are spoken.

I was never taught Navajo at home, so I was eager to learn what I could from my all–Native American high school and my Introductory Navajo Language class. That's where I heard this phrase for the first time. Although this was the first time I heard it in this language, I immediately recognized the importance of this teaching to my culture and my family.

This phrase is part of the Diné Protection Way Teachings, which are supposed to help people avoid certain acts and mindsets so that they can live a healthy and sacred life. That, however, is not how I learned this cultural lesson.

I grew up outside the Navajo reservation in New Mexico with my young biracial parents—my father, Navajo and white, and my mother, Black and white. Both started out working in fast-food restaurants and eventually completed degrees to become a police officer and a nurse. I grew up watching my parents constantly studying and working: one usually working the day shift, the other working nights, leaving my sister and me with our grandparents or aunts and uncles.

I was never explicitly taught anything about Navajo culture as a young child. My family, wanting the best for me, considered it easier to lean into my "white" side so that I wouldn't have to experience the racism and oppression they were subjected to. So instead they taught a traditional lesson of "never being lazy" through example and in the stories they told about the world around me.

That alcoholic cousin of mine? According to my family, he simply had too much time on his hands. My aunt in an emotionally abusive relationship? She is always sad because she didn't go to college to have a "real" career. Strangers approach you requesting money? Be sure to never give them anything; they would just spend it on alcohol and are only on the street because they would rather party than make an honest living.

In hindsight, it is clear to me that the common thread in those around me was not pure laziness. *It was trauma.* Trauma looms large in the Navajo community. Although the results of past suffering enveloped me growing up, we didn't talk about it. Instead, we focused on how we should move forward, how to be busy and stay productive.

As a child I felt that to be a *good* person I had to stay busy, so I quickly adopted this teaching as my life motto: constantly getting ahead in schoolwork, participating in as many clubs and activities as possible . . . reading, writing, practicing . . . anything so that I would never be sitting still. I felt a personal responsibility to avoid being a "drunken savage" or just another "Indian welfare queen."

While this hard work eventually got me into an Ivy League school to obtain my bachelor's degree in neuroscience and behavior, it did have negative effects. After years of never letting myself rest and pushing myself to always go the extra mile, my body finally forced me to slow down. I suddenly couldn't focus in my classes, and most days I struggled to get out of bed.

I finally realized that if I didn't get help I would risk dropping out of college. I eventually met with a therapist, who diagnosed me with Major Depressive Disorder and General Anxiety Disorder. I felt numb and ashamed. My family's judgmental words echoed in my mind. I felt lazy, that I should have been able to just push through. It took some time to accept that I was not just becoming complacent now that I had reached one of my long-term goals.

After receiving treatment and going to therapy for a while, I started volunteering in a lab and finally got a position in a research lab studying what I was most passionate about—the neuroscience of trauma. There I toed the line between success and burnout yet again. Frankly, I was still not fully convinced that the whole thing wasn't just in my head. *Doo hwił hóyéeda.* I didn't want to be lazy.

Working in a lab environment was intellectually, physically, and emotionally taxing. At first, I ran up and down the stairs between research trials, trying to keep my breath steady. I couldn't let it seem that being fat was slowing me down. If I ever wanted to be seen as a serious, successful academic, I felt I had to work twice as hard and never be caught finding things physically difficult.

Reading paper after paper about the visceral impact of trauma on the brain was also sometimes triggering and induced what my therapist lovingly called "trauma-flooding." I loved my work, but sometimes I felt that I was drowning in it. A typical busy day in the lab also involved analyzing last week's data and meeting with labmates to plan next month's project.

Much of my work revolved around experiments in which I observed how mice operate in stressful conditions. But sometimes watching how the brain managed stress felt too close to home. As a trauma researcher, I was studying brains at the same time I was trying to deal with my own—and that could be overwhelming.

I found myself asking, was I approaching burnout again? I had teamed up with a therapist to help me talk about and reflect on some of the complications of the ways I was taught growing up. At first the Diné teachings cut me like shattered glass because "don't be lazy" is a powerful lesson. Yet it is an incomplete one. When you're struggling with mental illness, people often say it is in your head. But the irony, which I discovered in part through reflecting in my hours of psychotherapy, is that neuroscience is also about what is in your head, and sometimes what is literally in your head matters the most.

I was drawn to neuroscience partly because it helped me understand that laziness wasn't the root of my community's struggles. Alcoholism, depression, and anxiety are not character flaws; they emerge from biological phenomena. In this sense, studying the neuroscience of trauma is a radical act of healing.

The lessons of my culture and my family, along with the science of the mind and a little talk therapy, little by little started to feel in harmony. The solution was not simple but rather involved a dizzying maze as I investigated the parts of my experience that didn't make sense. Psychotherapy gave me the space to reconcile the disparate parts. I would ask why it was so hard to just push through, and my therapist would gently say, "How can you succeed without caring for yourself?"

Neuroscience seeks to answer those questions in a way that doesn't dismiss the culture of my family or the teachings of my people. After a while, I didn't feel quite so raw and rundown. In fact, my principal investigator began working with me to create more flexible timelines, and we had regular check-ins to accommodate me; she has been a champion for my career and personal development. I was even given more opportunities to talk about my mixed heritage and make it a part of the culture of our lab. Now my identity is like a celebrated pastiche of a stained-glass window.

Living at the intersection of a multitude of identities certainly doesn't alleviate the pressure to be as perfect as possible. Woman. Native. Fat. Poor. Mixed-race. Mentally ill. It is difficult to be any of those things, but sometimes it feels impossible to be all of those things at once. The history of being Indigenous is the history of being forced to be more than one thing, of constantly being moved from location to location, from identity to identity. Paradoxically, I've found that, over time, my ongoing health and wellness journey has been a success *because* of these complexities, not in spite of them. Being an Indigenous person in the twenty-first century is a mixed experience because of the history of colonialism, genocide, and turmoil. Being Indigenous means always searching for stability of self, mind, and place.

To be Indigenous is not just to be some historical and mystical entity, but to be a quarter Black, a quarter Navajo, a daughter of New Mexico, and a New Yorker all at once. Perhaps that's why science and the pursuit of knowledge, both inside and outside the lab, have always made me feel whole. These days, I feel that I have the wisdom of the scientific method and the wisdom of my ancestors as a wind at my back.

Gabi Serrato Marks

30

THE RIDGE

GABI SERRATO MARKS

Content warning: n/a

I'm kneeling down in a mixture of mud and water, trying to stay as still as possible in a cave chamber I barely fit into. There are wet, sparkly rock walls on either side of me, and there are dozens of fragile stalactites—cave formations that grow from the ceiling—just a few inches above my back. I'm wet and a little cold, but it's my favorite room in this cave. I'm taking water samples and trying not to break any of the delicate formations, even though I feel like a moose trying to walk "carefully" in a thick forest. Water from a stalactite is dripping slowly into my tiny glass vial, so I use this time to catch my breath and take in the beautiful scenery, committing it to memory because phone pictures never come close to capturing the best parts of the cave. I only need about a tablespoon of water, but it's a slow drip. This part of my research doesn't take any strength; it just takes some patience and wiggling. I could do this every day if I needed to.

My small team is almost done with our work, so it's almost time to go back up to the surface. I'm dreading it, not because I don't want to leave the cave—I'm ready for a shower and quesadillas for dinner—but because I know that getting out will be even harder than coming in, and I am already exhausted.

The last water sample is almost full. My arm is resting on my knee and my shoulders are cramped, and I'm not quite sure how I'm going to unfold myself and get ready to head out. Everyone is tired, but I have Ehlers-Danlos syndrome (EDS), which means that my collagen doesn't work correctly. You can buy collagen in most grocery stores in the United States, but it's the raw ingredient for a recipe that my body never makes correctly.

No one comes to this cave for fun. There's a highway from San Luis Potosí, Mexico, to this small town, but once we get off the road there is no paved path to the cave. To get to this study site, we have to climb down hundreds of rungs on metal ladders that are anchored into the cave wall, courtesy of a mining company that pumps water out of the cave. When I get to a rickety spot on the metal ladders, I think about my grandfather, a retired ironworker, and put my faith in the welders who installed these ladders.

The cave is so huge that I've seen only a small portion of it. I don't know where the ladders end; there are pits so deep that even a headlamp with full batteries can't light them up.

EDS is a connective tissue disease, so every joint hurts me on a good day. I'm in rough shape after a few hours of caving and research, plus a few hours in the sun while we changed flat tires on our way here. I have a sunburn on my back, except for an X that is covered with thick exercise tape.

Before I became disabled, this whole day would have been enjoyable. I used to go hiking for fun. I willingly did a hike in New Hampshire that was supposed to take seven hours, and it actually took ten because I got lost and loved it. I played college water polo and went to practice five days a week. That's how I met my husband. Now I can only go to the grocery store if I clear my schedule for the rest of the day. My husband cuts up apples for me because it hurts my wrist to do it myself and it hurts my jaw—sometimes for days—to eat an apple straight off the core.

I need a whole week after fieldwork just to recover. I dread each day of caving, knowing that my joints and muscles will feel like I got hit by a rock hammer.

But there's no other way to get these water samples, so that's how I ended up here, deep in a cave, running on adrenaline, Diet Coke, and lime-flavored Fritos. There are a lot of perks to doing fieldwork in northern Mexico, and those chips are definitely one of them.

Once we finish our water collection, we're close to being done with the day's work, but we still have to get home. On our way out of the cave, we reach a slippery ridge. I knew it was coming, but I was still hoping that there would be a way around it.

There is mud caked into the treads of my hiking boots, so I struggle to get to the top of the hill where the rest of the team is taking a water break. I clamber over the rocks, not looking up at the cave ceiling because I know it always gives me a nervous chill. These "boulders" are here for us to climb because at some point, hundreds or thousands of years ago, they fell off the ceiling. So I don't look up; I just keep stepping onto the collapsed rock. Looking up also takes extra energy, and I need to focus on where my feet are going.

When I finally get to the top and see the cavernous pit to the right, swallowing all of our light beams, I start swearing under my breath. I have been thinking about this part of the trip for months, and somehow it's worse than I remember it. Looking at the next part of the path and the experienced cavers ahead of me, I see that I'll have to scramble—very carefully—from rock to rock to get to a lower level. From the top, it looks like I'm supposed to walk down a small waterfall, but without any water and with a thin layer of slippery mud on every surface.

It will only take a minute or two to get down the slope, it's barely any distance to cover, and there are people ahead of me and behind me to help. But it's the what-ifs that make me nervous because I'm a very clumsy caver. I can only partly blame EDS for that.

What if I slip off that one foothold and slide down, uncontrolled? I did that in the Yucatan, and my hand still has the scar to prove it. What if I grab a rock that looks solid, only to find that it's loose? I've done that more times than I can count. If it happens here, I'll slide right past the path I'm aiming for and down the dark slope into the pit. Some ancient cave formations would probably stop my fall.

Unfortunately, we're on our way out, so either I make it past this spot or I stay in the cave forever. I honestly considered that possibility.

I watch another grad student get through the slope, making mental notes about the rocks he uses as handholds and how he shifts his weight. I put my fear into a mental box because fear is dangerous down here. Fear makes you second-guess yourself. Before I can think about it too much or forget what he did, I sit down on the mud and slide my gear bag to him. It's my turn to face the slope.

But when I take a deep breath, I realize that, as nervous as I am, I'm having fun. Of course, if anyone had asked, I would have said I was having a great time during every trip. But I usually just tolerate this part of fieldwork. I never enjoy it.

So I'm a little surprised to check in with myself and find a smile. I wish I could say that I'm not tired anymore or can't feel the pain because I am just so focused on getting through this. But that's not true; I am sweating and cold at the same time, and I am aware of every bone and ligament from my feet to my neck because they all hurt.

Somehow, through a little bit of brain chemistry, positive thinking, or just good luck, I'm enjoying sliding down this muddy slope. It helps that I didn't tumble down into the dark, potentially bottomless pit. I will absolutely skip the ridge next time, but a few years of being disabled have shown me that it is possible to both hurt and have fun, to struggle and smile and really mean both.

I think back to that moment on the ridge whenever I get to choose between comfort and excitement. Even being able to make a choice is a privilege because sometimes my body decides for me. On those days, I have to lie on the couch, ice my joints, and do my best to get some work done. I'm not comfortable or having fun. The days when I hurt all over but can still walk to get an iced coffee and see Boston harbor are my good days.

When I feel torn between saving my energy at home or spending energy on people or experiences that I love, I remember that I still had fun at the bottom of a cave on a bunch of muddy rocks and that it was worth it. I'm still disabled, I still have chronic pain, but I also chase that joy whenever I can.

Vincent Martin

31

WHO I AM

VINCENT MARTIN

Content warning: grief, parent loss

THE SCHOLAR

During the summer of 2016, I trekked from the Atlanta area to visit my parents in Hogansville, Georgia. Visiting my hometown is always refreshing because it is a quiet town of approximately 2,700 people. I gave my parents a copy of my master's degree in human-computer interaction (HCI) from Georgia Tech. While they had attended my graduation ceremony in May of 2013, I had moved into a PhD program at Georgia Tech, and in 2016 I finally filed the paperwork to finish my 2013 master's degree to clear the rolls of the School of Interactive Computing.

What was unique about the copy of my degree was its $250 diploma frame. I felt that the degree deserved such an expensive frame because it took over thirty years to finally receive it. My father immediately took it down the hall and mounted it on one of the walls of the den that had held many of their children's academic honors from over the years. My father loved the frame and degree, and my mother squeezed my hand and said, "Good work, son."

When my mother and father attended my 2010 graduation for my undergraduate engineering psychology degree, she said,

"I'm not coming here for any more graduations. Quit playing." I knew what she meant was that I needed to finish what I had started in 1982.

I told my mother that she would not have to travel all the way to Marietta, Georgia, for any more graduations—she could stop at Georgia Tech in Atlanta for the rest of them. I was accepted into the master's program in HCI and eventually put that degree in the gorgeous frame hanging on my parents' wall. Georgia Tech was also the school I had first attended as an undergraduate and left in early 1982.

It was in the 1980s at Georgia Tech when I began having peculiar vision issues that affected my ability to study and keep my grades up. I decided to transfer to get a break. Even though I suspected that my vision was the problem, I also felt that I wasn't good enough to be at Georgia Tech; I felt broken. It's something I could never admit to my parents. But my mother's squeeze when I gave them the copy of my master's degree extinguished those feelings of inadequacy. I had returned and accomplished even more without any eyesight, becoming the first totally blind graduate in the school's 131-year history.

THE PARALYMPIAN

While I have always been a scholar, I have also always been a competitive athlete; I don't just play when it comes to sports.

When I started training for the 1996 Paralympic Games in Atlanta, Georgia, I realized that it was the first time I had found people like me—other competitive blind athletes. We had the same competitive attitude and treated one another as equal competitors.

During the Paralympics, I participated in athletics, better known as track and field. In 1996 there were twelve of us with some form of visual impairment on the fifty-plus athletics team. I was struck by the fact that everyone was also competitive and accomplished outside of sports—we also had multiple advanced degrees and were employed. It was completely contrary to the statistic saying that 70 percent of visually impaired people are unemployed.

The Paralympics represented a group of motivated and accomplished individuals, not just in sports but also in various aspects of their lives, often in education. We spoke to one another like regular people, and we were all satisfied with our own ability or disability. No one was jealous of someone else's situation; we were all "comfortable in our own skin."

Coming into the Olympic stadium in Atlanta for the games was one of the highest points of my life. Walking through the tunnel under the stadium, I heard a choir of five thousand singing. One of the singers ran up to me and said, "I'm your cousin!" and then said she'd see me at the family reunion afterward. I was taken aback by the excitement in her voice. The stadium was filled with eighty-five thousand people watching us during the opening ceremonies, which were grand and amazing by themselves, never mind the athletic events. A friend leaned over after the opening ceremony and said, "Well, I guess we can go home now."

When asked which events I participated in as a Paralympian, I used to say, "all of them." In the decathlon, a total of ten events over two days, the winner gets the title of "world's greatest athlete." The same applies to multi-events in track and field that have seven (septa-) and five (penta-) components. Able-bodied Olympic athletes, master's-level athletes, and many disability-related groups all compete in the pentathlon. During the 1996

Paralympics in Atlanta, Georgia, I participated in the pentathlon, the 100- and 400-meter relay races, and the open 100-meter dash and discus throw. My fifth-place finish was hindered by a fluke disqualification.

At the end of the games, my mother came up to me and said, "I see you've achieved your goal at the Paralympics." When I said, "I didn't get a medal," she said, "Your brother is here with two hundred of his friends screaming for you, your dad is wearing the same jacket as you right now . . . they know you're OK." I said that I knew I was OK before, that I was a normal person, and she replied that my family hadn't known that but that now they did. She said, "Well, I'll be in Sydney in 2000, and it would behoove you to show up." My mother knew that the Paralympics allowed me to maintain the level of competition I needed to keep working and training full-time.

In 2000 I traveled all the way to Sydney, Australia, with my mother. I made the discus throw my marquee event for 2000 and for the following Paralympics in Athens, Greece, in 2004. When I retired after 2004, I was the top-ranked thrower for the totally blind classification for the United States Association of Blind Athletes and had the distinction of making the second-farthest throw in history for totally blind athletes. The lifelong relationships I developed during my time as a Paralympian are with people all around the world. I also made connections between the athletic world of disability and my own goals as a scholar, becoming a rehabilitation engineer as part of my long career.

FROM A FAMILY OF PROBLEM SOLVERS

At the end of both my parents' lives, I held their hands and told their unconscious forms how much I loved them and most of all how much I thanked them for their assistance. The three times I represented the United States at the Paralympic Games and my athletic records were just a way of "making my father proud" of me again. My mother, the Rev. Bradell Allison Martin, passed away in March 2017, and my father, Charlie F. Martin, passed away in June 2020.

In their fifty-six-year marriage, they saw many changes, including the signing of many civil rights laws in the 1960s. The Fair Housing Act of 1968 allowed them to purchase their first home in 1969. My mother, an African American woman who could not afford to attend college although she graduated as the valedictorian of her high school in 1958, and my father, a man with an eighth-grade education, became solid members of their community and instilled values into all of their children that still resonate today. My 2016 degree from Georgia Tech was the culmination of many years of hard work and was the crowning achievement of the type of family they wanted to build.

"Quit playing" is exactly what my mother would say, and my father would nod his head in agreement. I added an additional master's degree from Georgia Tech, which made me the first and second graduate as a totally blind person. That degree, which became my seventh one, is a partial personification of the person that I became from the assistance and love from my parents; they taught me to pay attention and to be able to solve any problem with the information I was given.

I became aware of my blindness in my senior year of college, after I transferred from Georgia Tech to Southern Polytechnic. I was watching a basketball game when I saw a commercial for the Retinitis Pigmentosa Foundation; the former owner of the Cleveland Cavaliers basketball team was blind as a result of the disease. I sat alone on the couch in my apartment and finally realized what was wrong with me: I was slowly going blind, and

I was the only one who could see it. I ran to the library that freezing Sunday afternoon in February without a jacket, excited to find evidence to support my idea. Less than thirty minutes later, I had six different encyclopedias open on a desk. I had a huge grin on my face as I read all the symptoms of this disease—retinitis pigmentosa. It all finally made sense: I was not going crazy; I was just going blind! My blindness does not define me; it is just the way that I perceive information with my eyes.

THE MENTOR

I have no biological kids, but I constantly pass along my experiences and knowledge to others. I have over thirty mentees now because they never "leave" me—as they mature, I just consider them as junior colleagues. Watching young adults mature into strong and capable adults is the ultimate testament to who I am as a person with a disability. Watching my niece mature into a phenomenal third-grade teacher was great, but realizing that she never introduced me as her "blind uncle" was even better. I am the person who smiles when a mentee comes into town, says, "I've got it," and then pays the bill when we leave the restaurant after lunch or dinner. It has happened over twenty times, and I always just say, "Thanks."

I am the person who told his latest biomedical engineering and premed mentee three years ago that she was doing too much and that she could not work anymore while going to school full-time. She was confused when I laughed after she said she needed the work-study job. The confusion turned into disbelief and exasperation when I told her that she would still get the money without doing that job. She finally paused long enough to realize that I had just replaced her work-study job with a stipend. It was satisfying to help her focus on her studies instead of worrying about a job to pay her way through school.

Over the past three years with this same mentee, each monthly financial allotment has seen the thanks reduced from a full paragraph to a few thankful sentences and finally to just the two wonderful words "thank you." When I heard her name called at her graduation in May 2021, I knew that she had walked across the stage. I was also saying "next" to myself as my mentee entered the next stage of her life and I welcomed the next mentee into mine.

I met my mentee's parents in June this last year at the funeral of a family member. Her parents were excited to finally meet me and thanked me for all the guidance I have given their daughter over the past four years. I replied that the gift of all the love, affection, and education I received from my parents during my lifetime is why I do what I do and who I am as a person. "My mother was a Methodist minister. If she were here, she would tell me, 'You don't get any credit for doing what you are supposed to do.'"

VINCENT

I am a totally blind Paralympian, a seven-degree-holding PhD adjunct college professor, a digital UX assistive technology analyst bank employee, a nonprofit board member, and a State of Georgia advisory board member.

If you get to meet me, be prepared to understand that the smile on my face is genuine and that my upbeat attitude is my natural state. I'll greet you with "My name is Vincent. What would you like me to call you?"

I'm still playing, too, but it's a different game now.

Michele Cooke

32

THE BEST PLACE FOR MY HEARING AIDS IS ON MY DESK

MICHELE COOKE

Content warning: n/a

The moment I get to my office desk, I take off my hearing aids. As I do so, the world around me softens. Gone are footsteps, hallway chit-chat, brushing clothes and papers. Now I can finally focus.

With my hearing aids, the world's sounds are harsh. High-frequency sounds that I can't usually hear are amplified by hearing aids. Hearing aids make all the frequencies of speech louder than my hearing thresholds, which vary with sound frequency: high-frequency sounds need to be extremely loud for me to hear them. For example, I don't hear birds or the difference between "M" and "N," and while I can hear some vacuum cleaner noise, this sound is soft and easy for me to ignore. I'm often amazed that these two tiny machines that fit behind my ears can amplify all these sounds. But the most amazing part of the hearing system is the brain. How does my brain process all the incoming acoustic information compressed into the narrow volume range of my hearing? I don't know. I'm a physical scientist and have no idea how my brain operates, but whatever my brain is doing must be very hard work. For example, distinguishing the "T" and "F" sounds from the background noise of fans or cars passing by, which is critical for understanding speech, drains my energy. After an hour of listening, I start to drift away from the discussion. I experience listening fatigue.

Because my job as a professor at a primarily hearing institution involves a lot of oral communication, I have to manage listening fatigue every day. Consequently, any time that I am not required to listen, I pull off my hearing aids and put them on the desk. Immediately, the world becomes more familiar. My shoulders relax—I am home. In the gentle world of my natural hearing, my energies can shift from the challenging chore of listening to other tasks that I enjoy. I no longer have to work to distinguish speech from background noise and carefully speech-read to fill in the gaps in my understanding. I can spend my energy on science. I can be my authentic deaf self.

I didn't always think of myself as deaf or disabled. Before graduate school, I used to say that I was "hearing impaired." This identity aligns with the medical model of disability, which focuses on the deficit within the individual's body or mind. I was diagnosed with sensorial-neural bilateral hearing loss in kindergarten when a savvy teacher noticed that I was speech-reading her. This was back in the 1970s, before routine early childhood hearing screening. The revelation of my hearing loss triggered many years of appointments with audiologists, surgeons, and speech therapists, who were paid to try to make me adapt to and function more easily within the hearing world. These medical professionals perceive our disabled bodies and minds as deficient, and so our medical diagnoses deliver to us our first identities based on the medical model of disability.

But what happens when we flip the medical model around? Instead of changing me to become more hearing, what if we change the environment to become less disabling? This is the social model of disability, which posits that environments can be disabling to people. I was introduced to this model of disability in graduate school when I met other deaf and disabled people. Through those transformative conversations, I learned to recognize disabling environments. Lectures in dark rooms are disabling to me. Noisy conference poster halls are disabling to me. This shift in thinking empowered me to let go of "hearing impaired" as my identity and embrace being part-deaf; the label "hard of hearing" has never sat well with me. To me, "hard of hearing" emphasizes the hard work that I need to do to pass as hearing, while "deaf" and "disabled" emphasize my perspective on life and on science.

This shift in disability identity also maps onto how I use assistive technology. Audiologists always instruct me to wear my hearing aids for all waking hours to train my brain to process the new sounds. They have tried scolding me for not wearing my hearing aids enough. The medical expectation that I be as hearing as possible sets me up for constant failure because I will never be hearing. Instead of striving to be hearing, I choose deafness whenever I can. At the same time, I choose to use technology when I need to interact with hearing people who don't know how to communicate with deaf folks. At work, I use hearing aids, assistive listening devices, and transcription software to make communication with hearing colleagues less disabling.

By taking my hearing aids out and putting them on my desk, I become my authentic self, who happens not to hear birds or sharp consonants. My authentic self enjoys thinking about how faults in the Earth's crust evolve over time. My authentic self enjoys setting up laboratory and numerical experiments where my graduate students and I test the impact of various processes on fault evolution. My authentic self enjoys coding and developing new software tools to explore fault system evolution. My authentic self loves working with graduate students to figure out why their laboratory or numerical models didn't produce the expected results. My authentic self also has terrible drawing skills and enjoys being in the woods. In my authentic world, I can focus and work: composing articulate emails, coding scripts to analyze data in new ways, and crafting papers or proposals. My hearing aids are tools to help me interact with people around me, but they are not part of my identity as a scientist or as a person.

I recently read Jillian Weise's piece "Common Cyborg" in Alice Wong's *Disability Visibility*, which refers to disabled people who use technology to navigate the world as *cyborgs*.[1] What an interesting idea! I grew up watching science fiction shows and admired cyborgs, such as the Bionic Woman and later Geordi from *Star Trek Next Generation*. In every episode of *The Bionic Woman*, Jaime Sommers saves the day with her badass bionic hearing skills. While my hearing aids don't give me badass hearing, they are similar to Geordi La Forge's visor, which gives him enough vision that he can operate as if he were sighted. Folks on the *Enterprise* don't act weird about his visor and value Geordi as a fully contributing member of the team. However, these television cyborg role models hardly ever remove their technology. They portray a message that our technology improves us. With technology we might be able to contribute to the team or save the day, but without technology we are far less valuable. This thinking arises from the medical model of disability.

My hearing aids are like my car; I appreciate being able to travel places without being out in the elements, but I don't love driving, and I'm not dependent on my car to go everywhere. I do take better care of my hearing aids than I do my car—I dress them up with blue glitter ear molds. On the few occasions when I've misplaced my hearing aids or run out of batteries, I feel that something's amiss, as if a helpful friend that I've come to rely on isn't there for me. This has happened before teaching a few times. Teaching is a time when modulating my speech and hearing student questions and comments is important—at least within a classroom of hearing students. I could cancel my class. I could announce, "The instructor is not *able* to teach today." But that isn't quite right. I am able to teach; it will just be different. Although I miss that dear friend, I can rely on myself.

In both instances that I ran out of batteries before class, I proceeded with my lectures and explained to the class that I would be deafer than usual that day. Was the science that I taught that day of lower quality? Did the equations that govern the physics of rock deformation lose their magical power of relating regional contraction to uplift of the Earth's crust? No. I still guided a discussion of the material even though my speech was probably blurry. The students still asked great questions, even if they had to repeat themselves and speak clearly (no mumbling!). In a way, isn't this the ideal social model of disability? Instead of expecting me to strive to be a hearing professor, why not expect that each of us makes the classroom less disabling to one another? In this way, we can stumble together toward universal design and inclusion.

Science, technology, engineering, and mathematics (STEM) fields valorize ability. We applaud accomplishments and grade perceived effort and outcomes. We often glorify long hours at the lab bench, at the computer terminal, and in the field. The ability to orally deliver clear and impactful treatises on esoteric processes is the hallmark of an accomplished scientist. When I realized that I would have to teach without my hearing aids, my own internalized ableism was appalled. How dare I expect students to accommodate my hearing impairment!? How dare I appear as a flawed instructor instead of perpetuating the image of the all-knowing and influential scientist!? By teaching without my hearing aids, I risk invalidating my role as an instructor. How dare I appear

Flawed

Weak

Deficient

Disabled

Human?

I reject the supposition that scientists need to be perceived as invulnerable. By disclosing our differences and vulnerabilities, we show students that people, rather than facts, are the heart of the STEM enterprise. Additionally, mainstream STEM culture can learn from disabled scientists. Our approaches, with or without cyborg enhancement, expand the narrow definition of who can be a scientist and expand the traditional ways to do science. For example, one of the best oral science presentations I ever attended was delivered by a colleague with a pronounced stutter. He used visuals in creative ways so that he didn't have to rely on his speech. I learned from him how to employ visuals in my own talks to make them understood by a wider audience. These crip hacks (life hacks, but for disabled folks) that we employ leverage our disabled experiences to make science more accessible. Not only does STEM need to include more disabled scientists among our ranks, but STEM also needs to pay attention to how we are radically changing the ways that we do science to make it less disabling.

Am I less of a scientist when my hearing aids rest on my desk? Without my hearing aids, I am an innovative scientist exploring the evolution of faults within the Earth's crust while also eroding the disabling nature of science.

NOTE

1. Jillian Weise, "Common Cyborg," in *Disability Visibility*, ed. Alice Wong (New York: Vintage, 2020), 63–74.

CONCLUSION

Aid to Navigation

SKYLAR BAYER AND GABI SERRATO MARKS

"Aid to navigation": any device or marker that assists navigators in determining their position or safe course or that warns them of dangers or obstructions ahead. Such aids commonly include lighthouses, buoys, fog signals, and day beacons.

We hope each chapter of *Uncharted* has given you a different perspective on a disability you live with, a window into the multifaceted lives of disabled scientists, or perhaps something more like validation. Our lives and stories are often hidden, so we hope that, by sharing a few of these stories from the disabled science, technology, engineering, and mathematics (STEM) community, we can help chart the path ahead for the next generation of disabled scientists.

As editors, we were surprised to see how many authors (including ourselves) had shared common experiences despite having vastly different diagnoses. We found that there were many ways that academia and research science can be unfriendly to both visibly and invisibly disabled people. Attitudes of peers, teachers, and institutions can make or break careers in science; this fact is made particularly clear by the authors who have decided to leave the field altogether. These authors are still very much connected to science and apply their scientific training to their careers, but many of them may also feel pushed out of STEM.

The intersection of disability with race, sexual orientation, gender, income, education, and other factors also compounds the stress and burdens that our community—and the authors in this book—have experienced in their personal and professional lives. Disability is the only marginalized identity that is not broadly recognized as a form of diversity, as a community, or as a strength. This problem frames the retaliation and alienation that some researchers fear because of the ableism that abounds in academia; several authors chose to contribute anonymously, and some did not—both for good reasons.

We were also struck by our different perspectives on different events or experiences—a reminder that we don't share identical viewpoints just because we've experienced the world with disabilities or medical conditions. We may use different language (for example, Gabi never uses "impairment" in the context of disability, but others do) or have dramatically different preferences for access. Some of us may proudly identify as Disabled, while others may be trying to return to being "healthy" or some other nondisabled state. We might be managing the symptoms of an unknown condition, uncharted but self-diagnosed, that academia will not recognize for accommodations. What ties us together is the creative problem-solving skills and persistence we have developed. These qualities make the authors in this collection (and disabled scientists more broadly) particularly well suited to tackle complex scientific questions. We can see confidence, advocacy, resilience, and community building. These stories highlight how we can come together to create strong support systems.

The Americans with Disabilities Act (ADA) was created to reduce barriers for disabled people in the United States, serving as a minimum standard for accommodations. This foundation of the accessibility puzzle is supposed to lead to a strong support system. When support or accommodations are missing, disabled people are forced to advocate for themselves while departments, advisors, principal investigators, and others stand against them. Being torn in multiple directions by advocacy, medical management of disabilities, research, and more can lead to burnout. When disabled people leave STEM because of discrimination or poor treatment, science suffers. True access involves thoughtful dedication to creating a culture of inclusion and understanding of all disabilities that allows everyone to perform at the highest level. Before the ADA, there may have been even fewer scientists with disabilities and medical conditions, or at least fewer who could be up-front about them. In fact, before a few decades ago, it may not have been possible to get any scientists to share their stories in this type of anthology. That said, past disabled scientists and organizations have paved the way for many of us over the years; many of these organizations can be found in the Further Reading and Resources section.

Both of us are hopeful that inclusion and access will improve in the future through justice-centered advocacy. But we know that hope takes us only part of the way. We need allies—both disabled and nondisabled—to reduce the impacts of ableism and create true access in STEM. Science and disability rights are both shaped by collaboration. By reading these stories, you have already started that advocacy.

Going forward, we ask you to educate yourself on the history of disability while reflecting on how people around you might experience barriers and successes similar to those highlighted in these stories. There's much you can do to make the scientific world more welcoming, from learning more to acting as allies. Thank you for joining us on this journey.

ACKNOWLEDGMENTS

As coeditors, we have many people to thank for the creation of *Uncharted*. We thank all the authors that pitched to us back in 2018 when we didn't know how this was going to come together, as well as all the authors we've approached since then who have been willing to throw their souls into this project—we have loved working with and learning about these wonderful people. Special thanks to author Syreeta Nolan for reviewing, editing, and contributing to our conclusion and Further Reading and Resources section and to contributors Olivia Bernard and Alma Schrage for helping us develop the glossary.

When we realized in 2019 that we needed to get our voices out ahead of any book project, Rebecca Boyle gave us important advice about which magazines to reach out to and whom to contact. Thank you, Rebecca. We also thank Michael Lemonick for being interested and willing to work with us on our first *Scientific American* article; it was a thrill to not only be published online but to also have a version appear in a print edition.

Thank you to Miranda Martin, our editor at Columbia University Press, who emailed Gabi and asked if she had any book ideas and Gabi was able to say, "Well, actually, we do!"

Miranda has graciously helped us through every step of the publishing process, a new and scary (and uncharted) journey for both of us!

We thank Jennifer Galvin, who suggested that we reach out to the Alfred P. Sloan Foundation to seek funding. We thank the Alfred P. Sloan Foundation for providing funding for all authors and for giving us additional funds to hire an illustrator, Tiffany Chen. We are grateful to be able to pay our contributors up front, as we believe that all disabled people deserve fair compensation for their work.

Skylar thanks her colleagues at the National Oceanic and Atmospheric Administration (NOAA), the Ecological Society of America, and Roger Williams University for being enthusiastic and supportive of the *Uncharted* project and for seeing this work as an important contribution to the science community.

Gabi appreciates the encouragement and support from peers and colleagues at the Massachusetts Institute of Technology, Massive Science, the Broad Institute, and Stellate Communications.

We thank The Story Collider, especially producers and friends Ari Daniel, Katie Wu, and Emma Young, for helping us craft our stories and improve our storytelling skills, and also for encouraging us as friends and fellow storytellers. We also thank Alice Wong, editor of *Disability Visibility*, and Ashley Juavinett, author of *So You Want to Be a Neuroscientist?*, for insights into the book-publishing process.

Our support systems have been critical to helping us accomplish this monumental task, from reading our drafts of proposals and essays to encouraging us and believing in us. Skylar thanks Thom for believing in her no matter what, Bob and Deb for their unconditional support as her proud parents, and Millie, Echo, Misha, and Killick for necessary walks and playtime. Gabi thanks Alex, Spock, and Moose and her extended family. We could not be here without your love and belief in us.

Finally, way back in 2018, Skylar performed a version of her story in *Uncharted* at a producers' retreat for The Story Collider. At that point, Gabi already had a story of hers on *The Story Collider* podcast, originally recorded in late 2017—before she even had a formal diagnosis. It was her first foray into storytelling. Erin Barker, a friend of Skylar's and the current executive director of The Story Collider, suggested that Skylar publish her story in print and also maybe go talk to Gabi, since Gabi had also recently told a story about fieldwork and disability. So, thanks to Erin Barker, the coeditors of this book got together and started writing a book proposal, something we had never done but were excited to try. Without a little push and a little support, we wouldn't be here, so thank you to all those (including Erin) who have believed in us and pushed us to dream a little bit bigger than we could imagine on our own.

REFLECTION AND DISCUSSION QUESTIONS

We hope that this anthology can serve as a resource for *everyone* to learn more about disability in science, but we also want it to be useful for formal courses about storytelling and disability, student groups, book clubs, and more. We've developed these questions to help guide discussions of each section of the book.

GENERAL

- What does the word "uncharted" mean to you? As you read each story, take notes on what is uncharted in it.
- How is navigating through the world of science uncertain? How does that navigation change if you also have to deal with accessibility challenges or medical conditions?
- What characteristics do you consider essential to being a scientist? Do these authors challenge any of your assumptions?
- How have these scientists been excluded or discouraged? How have they been supported?
- How does intersectionality impact resources and belonging in scientific institutions? Note that most of the authors are women, and more than 40 percent are LGBTQ+.

- Did you learn about any new medical conditions or disabilities? If so, describe what was new information and identify what information you may want to look up to learn more.
- After reading this book, list five ways we could improve scientific culture to make science more accessible to those with medical conditions and disabilities.

PART I. GETTING UNDERWAY

- What were the most important events in these stories, and how did they affect the authors' long-term relationships with science?
- How does being different affect these authors' ability to be a researcher? A student? Do they consider themselves different from their peers?
- Sami's letter to her cousin differs from other chapters. Why is it important to offer support and advice to a relative who has a similar disability? Is this letter similar to any supportive letters or conversations you've had? Explain any similarities or differences.
- Identify an important event in your life that helped you "get underway" with your path. Explain why it was important. Are there any parallels with the stories you've just read?

PART II. BETWEEN THE DEVIL AND THE DEEP BLUE SEA

- What are the "devil" and the "deep blue sea" that the authors are trapped between in each story? Some stories may be clearer than others.
- What is certain in the authors' stories? What is unknown?
- How do these authors cope with difficult days?
- Hope and grief are prevalent in this section. How do patterns of emotional highs and lows differ among these stories?
- What particular stories in this section do you relate to? Why?

PART III. RALLYING THE CREW

- How do the authors receive and provide support to others in their stories?
- What kinds of help do the authors need?
- Are any of these authors reluctant to ask for assistance? If so, why do you think that might be?
- What kinds of help come from the following sources?
  - Family
  - Friends
  - Professional colleagues
  - Community and society
  - Institutions
  - Law, government, or policy
- What kind of support have you received from your "crew" to do well in school, science, or any other field? Do you notice any similarities or differences between your experiences and those of the authors?

PART IV. IN THE HEART OF THE MAELSTROM

- What is the "maelstrom" in each story? Do the authors exit the maelstrom? If not, what do you think the path forward is?
- How do you feel as you read these stories?
- What elements of the authors' experiences surprised you?
- Do any of these challenges feel familiar to you? Identify any "maelstroms" you've experienced in your life.
- Why do you think these experiences were so life-changing for the authors?
- How might these events affect other areas of their life?

PART V. REFLECTIONS IN THE WATER

- Describe the different ways these authors reflect on the events in their life. How have their views affected their broader outlook and perspective?
- What different kinds of coping mechanisms do the authors use to handle long-term struggles with diagnoses or multiple medical events?
- How do the authors' identities as scientists change in these pieces, if at all?
- Describe past events in your life that you view a little differently now than when they first happened. How has your perspective shifted?
- Why are some of the authors experiencing grief in these stories? What are they hopeful about for their future?
- What remains certain for the authors in these pieces? What remains uncertain?
- What is the social model of disability? How does it influence your thinking about these stories?

PART VI. I AM THE CAPTAIN OF MY SHIP

- How do the authors portray that they are the "captain of their ship"?
- Where does self-confidence come from for each of the authors?
- What kind of support do you imagine these authors have from their communities? Identify some specific examples of support you think they may have based on reading their stories.
- What advice do you think these authors would give to those in the "Getting Underway" section or other early career scientists?
- Have you ever felt that you were the captain of your ship? Describe the feeling, and explain why you felt that way.

FURTHER READING
AND RESOURCES

ARTICLES: NEWS, OPINION, GUIDES

Anti-Defamation League. "A Brief History of the Disability Rights Movement." Anti-Defamation League, 2018. https://www.adl.org/education/resources/backgrounders/disability-rights-movement.

Armstrong, Eleanor S., Divya Persaud, and Christopher A. L. Jackson. "Redefining the Scientific Conference to Be More Inclusive." Physics World, October 1, 2020. https://physicsworld.com/a/redefining-the-scientific-conference/.

Bernard, Olivia. "Despite the ADA, Science Often Isn't Accessible for Disabled People." Massive Science, August 2021. https://massivesci.com/notes/americans-with-disabilities-act-science-accessibility/.

Brown, Lydia X. Z. "Ableism/Language." Autistic Hoya (blog), November 2021. https://www.autistichoya.com/p/ableist-words-and-terms-to-avoid.html?m=1.

Burgstahler, Sheryl. "Making Science Labs Accessible to Students with Disabilities." University of Washington, 2012. https://www.washington.edu/doit/making-science-labs-accessible-students-disabilities.

Burke, Lilah. "Could Disability Be Further Included in Diversity Efforts?" Inside Higher Ed, November 2020. https://www.insidehighered.com/news/2020/11/12/could-disability-be-further-included-diversity-efforts.

Disability Rights Oregon. "Black People with Disabilities Are More Likely to Be Killed by Police Than White People with Disabilities." Disability Rights Oregon, June 2020. https://www.droregon.org/advocacy/black-people-with-disabilities-are-more-likely-to-be-killed-by-police-than-white-people-with-disabilities-1.

Gupta, Manisha. "Like Millions of Women, I Live with Chronic Pain—and I've Had to Learn to Advocate for Myself." Healthy Women, September 2022. https://www.healthywomen.org/your-health/advocate-for-myself.

Kelly, Shannon. "8 Influential Black Women with Disabilities to Follow." Disability Horizons, June 2020. https://disabilityhorizons.com/2020/06/8 -influential-black-women-with-disabilities/.

Lawrence, Anya. "Six Simple Steps to Make Fieldwork More Accessible and Inclusive." Geoscience for the Future (blog), October 2020. https:// geoscienceforthefuture.com/six-simple-steps-to-make-fieldwork-more -accessible-and-inclusive/.

Leary, Alaina. "How to Make Your Virtual Meetings and Events Accessible to the Disability Community." Rooted in Rights (blog), April 2020. https:// rootedinrights.org/how-to-make-your-virtual-meetings-and-events -accessible-to-the-disability-community/.

Meldon, Perri. "Disability History: The Disability Rights Movement." U.S. National Park Service, December 2019. https://www.nps.gov/articles /disabilityhistoryrightsmovement.htm.

Nolan, Syreeta. "Science, Accessibility, and Advocacy." Association for Women in Science 53 (Fall 2021): 42–43.

Schwarber, Adria. "ADA at 30: Scientists Urge Efforts Beyond Compliance." Physics Today, August 2020. https://physicstoday.scitation.org/do/10.1063 /PT.6.2.20200803a/full/.

Untonuggan. "Trigger Warnings 101: A Beginner's Guide." Medium, December 2017. https://medium.com/@UntoNuggan/trigger-warnings-101 -a-beginners-guide-e9fc90c6ba0a.

Vasquez, Krystal. "Excluded from the Lab." Chemistry World, December 3, 2020. https://www.chemistryworld.com/opinion/disabled-scientists-excluded -from-the-lab/4012695.article.

Vilfranc, Chrystelle. "#RevealToHeal: Mental Health and Communities of Color." Vanguard STEM, August 2017. https://conversations.vanguard stem.com/revealtoheal-mental-health-and-communities-of-color -4c25a62de69.

Vogler, Christian. "Accessibility Tips for a Better Zoom/Virtual Meeting Experience." Deaf/Hard of Hearing Technology Rehabilitation Engi- neering Research Center, March 2020. https://www.deafhhtech.org/rerc /accessible-virtual-meeting-tips/.

Weissman, Sara. "Protecting Students Who Seek Mental Health Treatment." *Inside Higher Ed*, August 2021. https://www.insidehighered.com/news /2021/08/12/justice-department-settles-brown-university-over-mental -health-leaves.

ARTICLES AND BOOK CHAPTERS

Bailey, Moya, and Izetta Autumn Mobley. "Work in the Intersections: A Black Feminist Disability Framework." *Gender and Society* 33, no. 1 (October 2018): 19–40. doi:10.1177/0891243218801523.

Braun, Derek C., M. Diane Clark, Amber E. Marchut, Caroline M. Solomon, Megan Majocha, Zachary Davenport, Raja S. Kushalnagar, Jason Listman, Peter C. Hauser, and Cara Gormally. "Welcoming Deaf Students into STEM: Recommendations for University Science Education." *CBE— Life Sciences Education* 17, no. 3 (2018). doi:10.1187/cbe.17-05-0081.

Canfield, Katherine N., Sunshine Menezes, Shayle B. Matsuda, Amelia Moore, Alycia N. Mosley Austin, Bryan M. Dewsbury, Mónica I. Feliú-Mójer, et al. "Science Communication Demands a Critical Approach That Centers Inclusion, Equity, and Intersectionality." *Frontiers in Communication* 5 (January 2020). doi:10.3389/fcomm.2020.00002.

Ellis, Katie M. "Breakdown Is Built into It: A Politics of Resilience in a Disabling World." *M/C Journal* 16, no. 5 (August 2013). doi:10.5204 /mcj.707.

Goring, Simon James, Kaitlin Stack Whitney, and Aerin L Jacob. "Accessibility Is Imperative for Inclusion." *Ecological Society of America, Frontiers in Ecology and the Environment* 16, no. 2 (March 2018): 63. doi:10.1002 /fee.1771.

Konrad, Annika M. "Access Fatigue: The Rhetorical Work of Disability in Everyday Life." *College English; Urbana* 83, no. 3 (January 2021): 179–99.

Nolan, Syreeta L. "The Compounded Burden of Being a Black and Disabled Student During the Age of COVID-19." *Disability and Society*, May 2021, 148–53. doi:10.1080/09687599.2021.1916889.

Sabatello, Maya. "A Short History of the International Disability Rights Movement." In *Human Rights and Disability Advocacy*, ed. Maya Sabatello and Marianne Schulze, 13–24. Philadelphia: University of Pennsylvania Press, 2014.

Serrato Marks, Gabriela, Caroline Solomon, and Kaitlin Stack Whitney. "Meeting Frameworks Must Be Even More Inclusive." *Nature Ecology and Evolution* 5, no. 5 (March 2021): 552. doi:10.1038/s41559-021-01437-9.

Sohn, Emily. "Ways to Make Meetings Accessible." *Nature* 576, no. 7787 (December 2019): 74–75. https://doi.org/10.1038/d41586-019-03852-2.

Solomon, Caroline. "An Often Overlooked Element of Diversity: Disability." *Limnology and Oceanography Bulletin* 30, no. 1 (2021): 20–21. doi:10.1002/lob.10425.

Woodcock, Kathryn, Meg J. Rohan, and Linda Campbell. "Equitable Representation of Deaf People in Mainstream Academia: Why Not?" *Higher Education* 53, no. 3 (2007): 359–79. doi:10.1007/s10734-005-2428-x.

BOOKS

Dolmage, Jay T. *Academic Ableism: Disability and Higher Education.* Ann Arbor: University of Michigan Press, 2017.

Girma, Haben. *Haben: The Deafblind Woman Who Conquered Harvard Law.* New York: Twelve Books, 2019.

Ladau, Emily. *Demystifying Disability: What to Know, What to Say, and How to Be an Ally.* Berkeley, CA: Ten Speed, 2021.

Pryal, Katie Rose Guest. *Life of the Mind Interrupted: Essays on Mental Health and Disability in Higher Education.* Chapel Hill, NC: Blue Crow, 2017.

Tobin, Thomas J., and Kirsten T. Behling. *Reach Everyone, Teach Everyone: Universal Design for Learning in Higher Education.* Morgantown: West Virginia University Press, 2018.

Wong, Alice, ed. *Disability Visibility: First-Person Stories from the Twenty-First Century.* New York: Vintage, 2020.

——. *Year of the Tiger: An Activist's Life.* New York: Penguin Random House, 2022.

VIDEOS, FILMS, AND PODCASTS

Engineering Change Podcast. https://engineeringchangepodcast.com/.

Crip Camp. https://cripcamp.com/.

7 Documentaries to Watch After *Crip Camp* (list). https://disabilityvisibility project.com/2020/05/04/7-documentaries-to-watch-after-crip-camp/.

TED. "I'm not your inspiration, thank you very much | Stella Young." *YouTube*, June 2014. Video, 9:16. https://www.youtube.com/watch?v=8K9Gg164Bsw.

SOCIAL MEDIA

#AcademicAbleism. https://twitter.com/hashtag/AcademicAbleism.

#DisabledInSTEM. https://twitter.com/hashtag/DisabledInSTEM.

Annie Tulkin (Accessible College). https://twitter.com/AcssCollege.

Britt DK Gratreak. https://twitter.com/BrittGratreak.

DARN (Disability Advocacy and Research Network). https://twitter.com /DARN_disability.

Disabled Academic Collective. https://twitter.com/DisabledAcadem.

Disabled in Grad School. https://twitter.com/DisInGradSchool.

Disabled in Higher Ed. https://twitter.com/DisInHigherEd.

Disabled Techie. https://twitter.com/DisabledTechie.

DREAM Disability. https://twitter.com/DREAMdisability.

Future Docs with Disabilities. https://twitter.com/Disabled_Docs.

Health Advocacy Summit. https://twitter.com/genpatient.

Sampson the Service Dog. https://twitter.com/sampson_dog.

Service Pup Basil Mae. https://twitter.com/servicepupbasil.

The ADHD Academic. https://twitter.com/theADHDacademic.

UC Access Now! https://twitter.com/AccessUc.

Voices of Academia. https://twitter.com/academicvoices.

ORGANIZATIONS

AAAS Entry Point! Program. https://www.aaas.org/programs/entry-point.

American Association of People with Disabilities. https://www.aapd.com/.

American Public Health Association Disability Section. https://www.apha. org/apha-communities/member-sections/disability-section.

Association of University Centers on Disability. https://www.aucd.org/template/index.cfm.

Association on Higher Education and Disability. https://www.ahead.org/.

Autistic Self Advocacy Network. https://autisticadvocacy.org/.

Black, Disabled, and Proud. https://www.blackdisabledandproud.org/.

Delta Alpha Pi Honor Society. http://deltaalphapihonorsociety.org.

DREAM College Disability. https://www.dreamcollegedisability.org.

Foundation for Science and Disability. http://stemd.org/.

Illimitable. http://illimitable.org/.

The International Association for Geoscience Diversity. https://theiagd.org/.

International Disability Alliance. https://www.internationaldisabilityalliance .org/.

The INvisible Project. https://invisibleproject.org/.

The National Alliance of Melanin Disabled Advocates (NAMD Advocates). https://withkeri.com/category/namd-advocates/.

National Alliance on Mental Illness. http://nami.org/.

Neighborhood Access. https://www.neighborhoodaccess.org/.

GLOSSARY

ABLEISM: Discrimination and prejudice against disabled people.

ACCELEROMETER: A tool that measures the rate of change in velocity of an object.

AGAR: A gummy substance made from red algae used for microbial culture.

ANTHROPOGENIC: Relating to or caused by human activity.

APHASIA: A disorder that affects the area of the brain responsible for language.

BILDUNGSROMAN: A type of novel that depicts the main character's formative years.

DEAF/DEAFBLIND/HARD OF HEARING: Several terms describing people from diverse cultures and communities with varying degrees of deafness who use a wide variety of ways to communicate, including any of the world's more than three hundred visual or protactile signed languages, many of the more than seven thousand spoken languages, and combinations of languages, augmentative and alternative communication, and writing. Degree of deafness and age of onset play a role in the term used, but a person's other identities (such as race, gender, sexuality, nationality, disability, and geographic location) also factor in what term a person will use.

DEAF (CAPITAL D): A term predominantly used in English-speaking countries that describes the community and its members who are deaf, deafblind and hard of hearing that share a culture and signed language created and passed down through social networks and schools. Deaf communities are notable for their history of educational, economic, and social achievement in face of efforts to suppress the use of signed languages. Many of these communities advocate for early sign language access and bilingual education for deaf and hard of hearing children to prevent developmental and educational delays caused by nonexistent or incomplete access to language. This occurs when language exposure exclusively relies on spoken languages, often under the assumption that assistive technologies like hearing aids or cochlear implants provide full exposure to spoken language.

DEFIBRILLATE: To stop the muscles of the heart from contracting using an electric shock to treat an irregular heart rhythm.

DISSERTATION: A written document on a particular subject, often required for doctorate degrees.

GASLIGHTING: Psychological manipulation that causes people to question themselves.

GEL ELECTROPHORESIS: The process of using an electrical current to push DNA fragments through porous gel to separate them by size.

GRAM STAINING: A laboratory method used to distinguish types of bacteria by staining their cell walls.

IMPOSTER SYNDROME: The experience of believing you are a fraud or not as competent as perceived.

INOCULATING LOOP: A tool used for transferring small microorganism samples.

LUMPECTOMY: Surgical removal of abnormal tissue, often used to remove cancerous tissue from the breast.

MIST NETTING: The use of baggy nylon or polyester netting to temporarily capture birds for research.

NEOTROPICS: Tropical regions experiencing rapid plant growth in parts of Central America, South America, and the Caribbean.

NEURODIVERGENT: People whose brains function outside societal expectations, including those with ADHD, autism, dyslexia, obsessive compulsive disorder, and other conditions.

OSSIFIED: Turned to bone.

RAISON D'ÊTRE: The reason for someone's existence.

RANGEFINDER: A tool used to measure the distance to a faraway object.

SALT PANNES: Areas of water retention and soft mud or clay in salt marshes.

SOLENOID: A wire coil that converts electrical currents into mechanical work.

SPECTROGRAM: A visualization of the frequencies of a sound.

SPECTROSCOPY: The study of atoms and molecules using light.

ABBREVIATIONS

ASL: American Sign Language

BIPOC: Black, Indigenous, and people of color

CART: Communication Access Realtime Translation

EMG: Electromyography

IUI: Intrauterine insemination

IVF: In vitro fertilization

LSAT: Law School Admission Test

MRI: Magnetic resonance imaging

NOAA: National Oceanic and Atmospheric Administration

PCR: Polymerase chain reaction

RNA: Ribonucleic acid

TA: Teaching assistant

USDA: U.S. Department of Agriculture

WAV: Waveform Audio File Format

BIBLIOGRAPHY

Avery, Erica. "Disabled Researchers Are Vital to the Strength of Science." *Scientific American Blog Network*, January 2019. https://blogs.scientific american.com/voices/disabled-researchers-are-vital-to-the-strength-of -science1/.

Baker, Mike. "Federal Agents Envelop Portland Protest, and City's Mayor, in Tear Gas." *New York Times*, July 23, 2020. https://www.nytimes.com /2020/07/23/us/portland-protest-tear-gas-mayor.html.

Bernard, Olivia. "Despite the ADA, Science Often Isn't Accessible for Disabled People." Massive Science, August 2021. https://massivesci.com/notes /americans-with-disabilities-act-science-accessibility/.

Bernstein, Maxine. "Portland Police Fatally Shoot Man in Lents Park." *Oregon Live*, April 16, 2021. https://www.oregonlive.com/portland/2021/04 /officers-responding-to-police-shooting-in-se-portland-park.html.

Blok, Vincent, Bart Gremmen, and Renate Wesselink. "Dealing with the Wicked Problem of Sustainability: The Role of Individual Virtuous Competence." *Business and Professional Ethics Journal* 34, no. 3 (Fall 2015): 297–327. doi:10.5840/bpej201621737.

Brown, Eryn. "Disability Awareness: The Fight for Accessibility." *Nature* 532, no. 7597 (April 2016): 137–39. doi:10.1038/nj7597-137a.

Dowling, Jennifer. "Tear Gas Used by Feds amid Protest Outside Hatfield Federal Courthouse." *KOIN*, March 2021. https://www.koin.com/news /protests/portland-protesters-protect-the-land-end-america/.

Everett, Glyn, Jessica Lamond, Anita T. Morzillo, Annie Marissa Matsler, and Faith Ka Shun Chan. "Delivering Green Streets: An Exploration of Changing Perceptions and Behaviours over Time Around Bioswales in

Portland, Oregon." *Journal of Flood Risk Management* 11 (December 2015): 973–85. doi:10.1111/jfr3.12225.

Everett, Glyn, Jessica Lamond, Anita T. Morzillo, Faith Ka Shun Chan, and Annie Marissa Matsler. "Sustainable Drainage Systems: Helping People Live with Water." *Proceedings of the Institution of Civil Engineers—Water Management* 169, no. 2 (April 2016): 94–104. doi:10.1680/wama.14.00076.

Flaherty, Colleen. "Federal Report Shines Light on Historically Underrepresented Groups in Science." Inside Higher Ed, May 2021. https://www.insidehighered.com/news/2021/05/04/federal-report-shines-light-historically-underrepresented-groups-science.

Lozano, Alicia Victoria. "Federal Agents, Portland Protesters in Standoff as Chaos Envelops Parts of City." *NBC News*, July 2020. https://www.nbcnews.com/news/us-news/federal-agents-portland-protesters-standoff-chaos-envelopes-portions-city-n1234520.

McRuer, Robert. *Crip Theory: Cultural Signs of Queerness and Disability.* New York: New York University Press, 2006.

Oliver, Michael. *Social Work with Disabled People.* London: Macmillan, 1983.

Renken, Elena. "How Stories Connect and Persuade Us: Unleashing the Brain Power of Narrative." *NPR*, April 2020. https://www.npr.org/sections/health-shots/2020/04/11/815573198/how-stories-connect-and-persuade-us-unleashing-the-brain-power-of-narrative.

Rittel, Horst W. J., and Melvin M. Webber. "Dilemmas in a General Theory of Planning." *Policy Sciences* 4, no. 2 (June 1973): 155–69. doi:10.1007/bf01405730.

Shanahan, Jessie. "Disability Is Not a Disqualification." *Science* 351, no. 6271 (January 2016): 418. doi:10.1126/science.351.6271.418.

Shapiro, Joseph. "She Owes Her Activism to a Brave Mom, the ADA and Chocolate Cake." *NPR*, July 2015. https://www.npr.org/sections/goatsandsoda/2015/07/31/428075935/she-owes-her-activism-to-a-brave-mom-the-ada-and-chocolate-cake.

Weise, Jillian. "Common Cyborg." In *Disability Visibility*, ed. Alice Wong, 63–74. New York: Vintage, 2020.

——. "Disability Impacts All of Us." Centers for Disease Control and Prevention, September 2020. https://www.cdc.gov/ncbddd/disabilityandhealth/infographic-disability-impacts-all.html.

——. "Prevalence of Disability and Disability Types." Centers for Disease Control and Prevention, October 2021. https://www.cdc.gov/ncbddd/disabilityandhealth/features/disability-prevalence-rural-urban.html.

ABOUT THE CONTRIBUTORS

PART I. GETTING UNDERWAY

Mpho Kgoadi is an astrophysicist and a PhD candidate at the University of the Witwatersrand. Mpho specializes in cosmology, specifically the role of dark matter in the early universe. He is living with transverse myelitis and is a wheelchair user. He is passionate about science communication, teaching, and motivational speaking. Mpho hopes to one day be a great researcher and teacher who can inspire the next generation of Black children in Africa to reach for their dreams. Mpho enjoys gaming and watching anime and superhero movies as well as reading novels and comic books. Social media: @mphodapoet on Twitter

Jenn Pickering spends half her time managing type 1 diabetes and the other half as an Earth scientist studying how rivers move sediment from mountain sources to the deep ocean. She did her MS and PhD work in South Asia, where she learned to speak Bangla, and has become an advocate for culture-conscious practices in scientific research, whether domestic or abroad. After spending more than five years working in research and development within the energy industry after her PhD, she made a career transition back to academia and is currently a senior research associate at the University of Kansas. Social media: @jenn_of_earth on Twitter; @jenn.of.earth on Instagram

Although **Maureen J. Hayden** grew up in the desert state of Arizona, she has become an advocate for marine life and our oceans. She is a PhD candidate at Texas A&M University and is working toward her doctorate in marine biology by studying plastic pollution on Texas beaches. She earned her bachelor's in marine biology from the University of Rhode Island in 2015 and her master's in biology from Walla Walla University in 2017. With an eye for innovative and multidisciplinary scientific studies, Maureen plans to make waves of change pursuing a career as a research scientist who studies the impacts of climate change on the marine environment. Social media: @maureen_marine on Twitter; @MaureenThe-MarineBiologist on Instagram

Sami Chen is a dyslexic aspiring Earth scientist in her sixth year of the Earth System Science PhD program at Stanford University.

Amanda Heidt (she/her) is a recovering marine scientist and current science journalist living in Moab, Utah. She received her master's in science communication from the University of California, Santa Cruz, and her master's in marine science from Moss Landing Marine Laboratories. Now a writer and editor at *The Scientist*, Amanda's past work has appeared in *Science, KQED Science, Discover Magazine*, the *San Jose Mercury News*, and the *Santa Cruz Sentinel*, among other publications. She misses the ocean, but her room has become a shrine to salty water in the desert, and she recently got a nudibranch tattoo. Social media: @Scatter_Cushion on Twitter and Instagram

PART II. BETWEEN THE DEVIL AND THE DEEP BLUE SEA

Daisy Shearer is a PhD candidate in experimental condensed matter physics at the University of Surrey's Advanced Technology Institute. Her PhD research focuses on semiconductor spintronics for quantum technology applications. Daisy holds an integrated master's degree in physics with first-class honours from the University of Surrey, where her master's research project involved working in research and development at the Centre for Integrated Photonics developing high-speed electroabsorption modulated lasers (EMLs) for telecommunications. She is a passionate researcher, science communicator, and educator with a drive to make STEM more accessible and inclusive, focusing on disabled and neurodivergent people because she herself is autistic. Social media: @QuantumDaisy on Twitter; @notesfromthephysicslab on Instagram

Lauren A. White is a quantitative disease ecologist and infectious disease modeler. Since finishing her PhD at the University of Minnesota in 2018, she has worked as a postdoc at the National Socio-Environmental Synthesis Center, served as an American Association for the Advancement of Science (AAAS) Science and Technology Policy fellow at the USAID Office of HIV/AIDS, and is currently working as a modeler statistician for the California Department of Public

Health's COVID Modeling Team. Beyond research, Lauren enjoys hiking, reading fiction, painting, and taking or teaching yoga classes. Social media: @LAWhite_ Ecology on Twitter; @noteanoprana on Instagram

Anonymous 1 is a first-generation immigrant and scientist trying to be a good person and leave the world (or her tiny part of it) a better place.

Skylar Bayer (she/her/hers) is a marine ecologist, storyteller, and science communicator who lives in Alaska. Her scientific research focuses on marine ecology, bivalves, aquaculture, and community engagement. She completed her PhD on the secret sex lives of scallops, a subject that landed her on *The Colbert Report* in 2013. She is an alumnus of the Sea Grant Knauss Marine Policy Fellowship and has been a producer for *The Story Collider* since 2014. When there isn't a pandemic going on, she also enjoys Brazilian jiu-jitsu, the gentle art. Social media: @drsrbayer on Twitter; @skylarrb26 on Instagram

Furaha Asani (she/her) is a public academic, mental health advocate, precarious migrant, and writer with a PhD in infection, immunity, and cardiovascular disease. Furaha is committed to the destigmatisation of mental illness and believes in bringing her whole self (flaws and all) to every encounter. Social media: @drfuraha_asani on Twitter

PART III. RALLYING THE CREW

Alma C. Schrage (last name pronounced SHRA-gee) is a pollinator ecologist currently finishing her MSc at the University of Illinois at Urbana-Champaign and doing pollinator research and assessments in the Chicago region. Deaf since birth, Alma uses American Sign

Language and English. She often uses her finely honed lipreading and guessing skills to understand the sign-impaired. Based on her experiences growing up in English-only settings, she believes in the importance of early sign language access for deaf and hard-of-hearing children of hearing parents. Alma talks like a book because that's where she learned most of her English, and she has a hard time writing dialogue. Her hearing aids are great for discreet Bluetooth music streaming at inaccessible conferences. Social media: @beebirdandbook on Twitter

When **Alexander G. Steele** isn't trying to figure out how the spinal cord works, he loves making things with his hands, but what really makes him happy is simply helping others. Overall, he's just a person trying to hold himself together long enough to make a positive change in the world. Social media: @LunaticLabs on Twitter

Sophie Fern is a writer and scientist who lives just down the road from the world's only land-based breeding colony of toroas, or northern royal albatrosses. She is fascinated with the natural world and will jump (not literally) at any chance of adventure. She is just finishing up a PhD study looking at nonhuman charisma and its role in the New Zealand conservation system, and her next projects are to get a job and to finally write a murder mystery set on Rangatira Island. Social media: @sophiefern on Twitter

Sophie Okolo, MPH, is a Forbes innovator and aging contributor, a TEDMED 2020 research scholar, and a Columbia University 2021 Age Boom fellow. Sophie is the founder of Global Health Aging, a creative consultancy and award-nominated resource featuring research, news stories, and diverse opinions regarding healthy longevity. For almost a decade, she has used her experience in health policy, scientific research, and patient advocacy to understand and highlight the importance of a life-course approach to aging. Her writing has appeared in *Salon, PBS Next Avenue,* and *Inverse,* among other outlets. Sophie has a BS in bioinformatics with research honors and an MPH in aging and health research. Because of her contributions to the field, she has been elected to the New York Academy of Medicine and the Sigma Xi Scientific Research Society. Social media: @sophieokolo on Twitter; @storieswithsoph on Instagram; Website: soinspiredhealth.com

Richard Wendell Mankin grew up near an air force base between two mountain ranges in southern New Mexico, an area full of technological and environmental diversity. He was led into science by his natural curiosity about cars, rockets, and starry nights on an Earth with deserts, mountains, and streams full of different animals and insects. After exploring different research labs in college and graduate school, he found a "sweet spot" that combined many of his interests: developing new technology and software that detect and help control insect pests hidden in food commodities, trees, and soil. Website: ars. usda.gov/rmankin

PART IV. IN THE HEART OF THE MAELSTROM

Anonymous 2 is a writer based in Connecticut.

Dr. Juniper L. Simonis (they/them/theirs) is a psychiatrically disabled freelance ecologist living in Portland, Oregon, who owns and operates DAPPER Stats and is the founding director of the Chemical Weapons

Research Center. When not doing science, Juniper is a four-time Women's Flat Track Derby Association world champion with the Rose City Rollers' Wheels of Justice. Social media: @juniperlsimonis on Twitter, Facebook, Instagram; Website: juniperlsimonis.com

Syreeta L. Nolan is a cofounder of Disabled in Higher Education on Twitter, a board member of Health Advocate X, and the founder of JADE (Justice, Advocacy, and Disability Education), a holistic disabled justice platform focused on empowering disabled students, faculty, staff, and alumni through community and support. She earned her bachelor's with honors in human health psychology from the University of California San Diego. She hopes to continue to obtain a PhD in health policy or prevention science toward her goal of transforming the mental health field through comprehensive preventive systems similar to those in the physical health system. Social media: @nolan_syreeta on Twitter

Amanda O'Brien is a PhD candidate living in the city with her partner and dog. She looks forward to the next time she can spend a week on the beach at the Outer Banks.

Stephanie Schroeder is the program manager for a National Science Foundation Engineering Research Center at the University of Minnesota. Trained as a marine biologist, she specializes in equity and inclusion in STEM. She received her BS from the University of Wisconsin–Milwaukee and her PhD from the University of Oregon. While investigating owl limpet territorial behavior, she realized that she loved integrating research, education, and social justice, an interest that was further reinforced when she was diagnosed with MS in 2015. She lives in Minneapolis and explores the city with her spouse and dog.

Divya M. Persaud (she/her) is a planetary geologist, writer, composer, and speaker. She is a postdoctoral scholar supporting missions to explore Jupiter's moon Europa, and she recently completed her PhD at University College London, where she applied 3D imagery to probe the geology of Mars. Her upcoming book, *do not perform this: a song cycle*, won the 2017 Editor's Choice Award from the "Great" Indian Poetry Collective, and her poems have appeared in *Anomaly*, *The Deaf Poets Society*, *The A3 Review*, and elsewhere. She has performed and spoken about her art, science, and advocacy internationally. Social media: @Divya_M_P on Twitter; @divyamper on Instagram; Website: divyampersaud.com

PART V. REFLECTIONS IN THE WATER

Glyn Everett is a sociologist studying architecture and the built environment (ABE) at the University of the West of England, Bristol. He researches public understanding, awareness, and preferences involving blue-green infrastructure used for flood risk management. He teaches the social model of disability and the importance of inclusive design practices to a range of ABE students and is keen to develop research in this area. Glyn has conducted research on using education and learning as tools of social empowerment, is a committed Gig Junkie, and is hungry to return to Portland, Oregon, and see how the bioswales are developing. Social media: @glynjamyn on Twitter, Facebook, Instagram

Emma Tung Corcoran recently completed her PhD in the Department of Molecular, Cellular, and Developmental Biology at Yale University. Her graduate work focused on studying epigenetic mechanisms that regulate genome stability, and she also dedicated substantial effort toward creating resources and programming to promote an inclusive scientific community at her institution. Outside of research, Emma loves reading, playing games, and baking. Emma currently resides in Massachusetts with her wife and their two cats. Social media: @emtungcorc on Twitter; @emtungcorc on Instagram

Leehi Yona (she/her) is a Mizrahi and first-generation climate researcher who is passionate about science, policy, and justice. She is a JD-PhD candidate at Stanford exploring the global carbon cycle and greenhouse gas inventories. A recipient of the Lieutenant Governor of Quebec Youth Medal, Leehi was named Canada's Top Environmentalist Under 25 and has written two books for young climate activists, informed by her experience as a youth organizer at United Nations climate negotiations. She holds a master's of environmental science from Yale and a bachelor's in biology and environmental studies from Dartmouth College. She loves painting and cooking for loved ones. Social media: @LeehiYona on Twitter; Website: leehiyona.com

Katie Harazin is an oceanographer from Atlanta, Georgia. She is finishing her PhD in past climate change and is also active in science communication and social media for professional scientific societies. In her free time, she enjoys digital art and illustration, and she secretly dreams of being an artist when she grows up. She currently resides in the U.S. South with a dog named Mr. Pickles.

Sunshine Menezes grew up off the grid in rural northern Michigan and later moved to Rhode Island to earn a PhD in oceanography. Unsatisfied with the ratio of time alone in dark rooms to time with others engaged in stimulating conversation, she left the lab to work on environmental policy and communication. She is a clinical professor of environmental communication at the University of Rhode Island (URI) and executive director of URI's Metcalf Institute. She cofounded the national Inclusive SciComm Symposium to build a community of practice around equitable, intersectional science communication. Social media: @SunshineMenezes on Twitter

Olivia Bernard is a disabled creative with a background in raccoon wrangling, shark chasing, and foraminifera flipping. She left research to create engaging experiences that explore our world and those imagined. She spends her free time reading speculative fiction for the science and reading geology textbooks to inspire fantasies. Olivia's favorite accommodation is a flexible work schedule (and snacks).

PART VI. I AM THE CAPTAIN OF MY SHIP

Jennifer L. Piatek is a planetary scientist (infrared spectroscopy, photometry, impact craters), college professor (geology and astronomy), and in general someone who is just trying to navigate the world from a wheelchair. Social media: @squawky on Twitter

Taylor Francisco is a neuroscience researcher at Mount Sinai and a graduate student at Columbia University in New York City. Her long-term goal is to increase the scientific understanding of trauma and to also become a physician-scientist advocating for these issues through public education, policy coordination, and patient care. As an Indigenous woman in STEM, Taylor strives to promote underrepresented youth in academia as well as systemic diversity and equity. When she's not in the lab, you can find her walking throughout the city trying all the vegan restaurants or watching squirrels in Central Park.

Gabi Serrato Marks is a geochemist turned writer. She received her PhD in the MIT-WHOI Joint Program in Oceanography and is now a partner at Stellate Communications, a public relations firm run by scientists for scientists. She can be found drinking iced coffee year-round in Boston with her husband and two cats. Social media: @gserratomarks on Twitter; Website: gabrielaserratomarks.com

Vincent Martin has had a long and diverse working background. In the past thirty years, he has worked in the field of rehabilitation in a number of capacities. He was initially a technology trainer, moved to rehabilitation engineering, and then worked as a rehabilitation research scientist for the U.S. Veterans Administration. He currently works as a digital user experience accessibility analyst for Regions Bank. He has attained seven degrees, including three undergraduate degrees, three master's degrees, and a dual doctorate degree in economics and systems engineering. He also is a retired Paralympian who competed in 1996, 2000, and 2004 in track and field. He is still the U.S.A. record holder in the discus throw and pentathlon in his category.

Michele Cooke is a professor of geosciences at the University of Massachusetts Amherst. Her research on earthquake and fault mechanics has earned her research awards, invited her to talks and fellowships, and placed her on the leadership of several professional organizations. While navigating her career with a disability, Michele has advocated for disability inclusion within academia and won awards in recognition of these efforts. She co-coordinates

The Mind Hears (www.theMindHears.org), a blog by and for deaf and hard-of-hearing academics that seeks to crowdsource strategies and create community among academics with deafness. Social media: @geomechCooke on Twitter; Website: theMindHears.com

A pangolin

ILLUSTRATOR

Tiffany Chen is a biologist seeking to work with pangolins and build bridges between diverse cultures. She has a long-standing interest in Chinese martial arts and has been practicing performance-based martial arts since 2013. Her interests lie in people and conservation. She has had experiences working with pigeons, bears, mountain lions, and other animal species and believes in the intertwinement of human cultural diversity and biodiversity. She hopes to continue in the fields of art, sciences, and martial arts as long as possible.

Printed and bound by CPI Group (UK) Ltd, Croydon, CR0 4YY

30/10/2024

14583732-0001